Nuclear Energy at the Crossroads

Books by Irene Kiefer

GLOBAL JIGSAW PUZZLE
The Story of Continental Drift

ENERGY FOR AMERICA

POISONED LAND
The Problem of Hazardous Waste

NUCLEAR ENERGY AT THE CROSSROADS

IRENE KIEFER

Nuclear Energy at the Crossroads

Illustrated with photographs

and with diagrams by JUDITH FAST

1984 ATHENEUM *New York*

LIBRARY OF CONGRESS CATALOGING IN PUBLICATION DATA

Kiefer, Irene.
Nuclear energy at the crossroads.

Includes index.
SUMMARY: Presents cases for and against the use
of nuclear power, considering its advantages
over other forms of energy as well as disadvantages
such as the problems of accidents, low-level radiation,
and radioactive waste disposal.
1. Atomic power—Juvenile literature. [1. Atomic
power. 2. Energy policy] I. Title.
TK9148.K56   333.79′24   82-1681
ISBN 0-689-30926-0   AACR2

Published simultaneously in Canada by
McClelland & Stewart, Ltd.
Composition by American–Stratford Graphic Services, Inc.
Brattleboro, Vermont
Printed and bound by
Fairfield, Graphics, Fairfield, Pennsylvania
Typography by Mary M. Ahern
Layouts by Marge Zaum
First Printing September 1982
Second Printing June 1984
Third Printing December 1984

Contents

Nuclear Energy
at the Crossroads

I.

The Nuclear Debate

IN THE 1950s, as the United States began operating its first nuclear power plant, some people hailed nuclear energy as *the* energy source of the future. One enthusiastic supporter even said that nuclear power would be so cheap that it wouldn't be worth metering, the way electricity usually is to determine how much a customer is using.

That claim certainly proved to be overly optimistic. Nuclear power, nevertheless, has become an important source of energy worldwide. In 1981, 78 nuclear power plants were licensed to operate in the United States; they generated about 12 percent of all electricity we consumed. In some parts of the country, the figure is much higher; Vermont, for example, gets about 80 percent of all its electricity from the atom. Plants now under construction could push the national average to about 15 percent by the end of the decade.

Some plants now on the drawing board may never be built, however. In recent years, plans for a number of plants have been delayed or canceled. In the period 1979–

The first commercial nuclear power plant began operating in Shippingport, Pennsylvania, in 1957. In 1981, 78 nuclear plants were generating about 12 percent of all electricity consumed in the United States. WESTINGHOUSE ATOMIC POWER DIVISION

1981, for example, plans for 28 plants were cancelled, there were delays in construction of 101 plants, and no new plans for building were announced.

There are several reasons for these delays and cancellations. One is that consumption of electricity has not grown as rapidly in recent years as had been expected. Higher energy costs have made consumers more aware of conserving power.

Another reason is that the cost of building nuclear

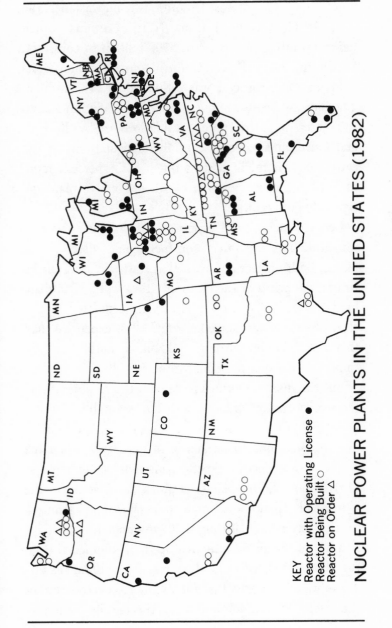

KEY
Reactor with Operating License ●
Reactor Being Built ○
Reactor on Order △

NUCLEAR POWER PLANTS IN THE UNITED STATES (1982)

power plants has increased enormously. For instance, the plant in Cowans Ford Dam, North Carolina, which began operating in 1981, cost $900 million to construct. When the Duke Power Company first announced its plans in 1969, the cost was estimated at $185 million. The increased costs are due primarily to inflation and to the increasingly strict safety regulations set by the Federal Government. Because of these regulations, in the early 1980s, it might cost as much as 25 percent more to build a nuclear power plant than a power plant designed to burn coal.

Because such a wide variety of factors, such as safety standards and the prices for fuel, could affect power plants, both nuclear and coal, that start operating in the future, the prices of the electricity they will generate are hard to predict. The electricity being generated by nuclear plants already built, however, is less expensive than that generated by burning coal, oil, or natural gas, although more expensive than hydroelectricity.

But the most important reason for the troubled times the nuclear power industry now faces is public concern about the safety, environmental, and health effects of its plants. That concern peaked in March 1979, when the accident that nuclear experts considered extremely unlikely almost did happen. One of the two reactors of the Three Mile Island power plant near Harrisburg, Pennsylvania, overheated, releasing small amounts of radioactive steam into the air over an area up to 32 kilometers from the plant. For a few days, there was great public anxiety. There were even fears that the highly radioactive core of the reactor would melt and break through its many pro-

An accident at the Three Mile Island power plant near
Harrisburg, Pennsylvania, triggered a national debate over the
future of nuclear power in the United States. One of the two
reactors there overheated, releasing small amounts of
radioactive steam into the air. The four large cooling towers
overshadow the rest of the plant. DEPARTMENT OF ENERGY

1. *Unit 1 Reactor*
2. *Unit 1 Control Service Building*
3. *Unit 2 Reactor*
4. *Unit 2 Control Service Building*
5. *Turbine Buildings*
6. *Fuel Handling Building*

7

tective "envelopes." In fact, there was only a very remote possibility that such a catastrophic "meltdown" would happen. Nevertheless, Three Mile Island brought to a head the debate over the role that nuclear power should play in the nation's energy future.

There have always been great fears about the safety of nuclear energy, in part because it was developed in secrecy and introduced to the world (in 1945) in the form of two powerful bombs that the United States dropped on Japan. But, in many ways, nuclear energy is a clean way of generating electricity. A nuclear plant requires less land, for example, than a power plant based on coal, taking into account the land used also in mining the coal or uranium and disposing of wastes. Nuclear plants do not release chemical pollutants into the atmosphere. Like all plants generating electricity from steam, a nuclear plant releases heat into the environment. Although a nuclear plant releases more heat than a nonnuclear plant of similar size, this can be controlled by installing cooling equipment or reusing the heat in some way.

It is the radiation produced in nuclear power plants that poses a grave threat if it escapes from the plant. You can't see it, smell it, hear it, taste it or touch it. But even in small doses, radiation may damage living organisms. If the damage is to the reproductive cells, it may be inherited by future generations.

The radiation produced by nuclear power plants poses three kinds of risks:

- An accident could release large amounts of radioactive materials.

- Day-to-day operations produce very small quantities of radioactive materials. The consequences of low levels of radiation—cancer and damage to reproductive cells, for instance—may not show up for decades. Furthermore, the effects of radiation are usually assumed to be proportional to the dose—that is, even small doses have some chance of producing such damage.
- Wastes from nuclear plants can release radiation for millions of years.

Opponents of nuclear power consider these risks so serious that no more nuclear plants should be built. They believe that the United States can produce energy for its growing population in other, safer ways.

Supporters of nuclear power point out, on the other hand, that nobody has been harmed by a nuclear power plant, and all the other ways of generating electricity also pose risks. Supporters of nuclear power believe that the risks are not great enough to stop building new nuclear plants. Moreover, our large natural resources could supply fuels for nuclear power well into the next century, helping to free us of our dependence on foreign oil, which, for political reasons, may not always be available.

The 1980s may well decide the future of nuclear power, perhaps the most complicated technical issue ever to come before the American people. Nuclear power also involves sensitive political, economic, and social issues. Only by understanding these issues can the people make wise choices in meeting their future energy needs.

The world was introduced to nuclear energy in 1945 when the United States dropped atomic bombs over two Japanese cities: Hiroshima and Nagasaki (this picture). DEPARTMENT OF ENERGY

WHERE IT ALL BEGAN

Nuclear power began in an unlikely place—in a make-shift laboratory in a squash court beneath the grandstands of the University of Chicago's athletic stadium. There a group of scientists and engineers were taking part during World War II in the supersecret Manhattan Project, a code name for a massive effort to develop an atomic bomb. The job of the Chicago group was to achieve a nuclear chain reaction. A chain reaction is similar to a fire. Like a fire, once it is started, it will burn as long as it is supplied with fuel. Experts in New York, Tennessee, New Mexico, California, and Washington were trying to solve other problems surrounding development of an atomic bomb.

On December 2, 1942, the Chicago group, led by Enrico Fermi, an Italian physicist, succeeded in starting and controlling a chain reaction in the world's first nuclear reactor. Though small and crude by modern standards, the Chicago reactor was the crucial first step that led to development of the atomic bomb—which undergoes an uncontrolled chain reaction when it explodes—and the development later of controlled nuclear reactions for peaceful purposes.

II.

How Radiation is Produced in Generating Nuclear Power

THE RADIATION produced by nuclear power origi-nates in the atom, the fundamental building block of the chemical elements. (The word atom comes from the Greek word *atomos*, meaning indivisible.) Molecules are groups of two or more atoms held together by chemical forces. All the substances we know are made up of either atoms or molecules.

The atom consists of a dense inner core, the nucleus. Except for ordinary hydrogen (the lightest of the ele-ments), the nucleus of all elements contains two kinds of particles: the neutron, which carries no electrical charge, and the proton, which is slightly smaller than the neutron. The proton carries a positive electrical charge and consequently the nucleus also has a positive charge.

(The nucleus of an atom of hydrogen consists of a single proton.)

Spinning in orbit around the nucleus at various distances are electrons, particles much smaller than neutrons or protons and carrying negative electrical charges. Electrons determine the chemical properties of the atom. Because opposite electrical charges attract, electrons are held in orbit around the positively-charged nucleus. Ordinarily, the negative charges in the electrons just balance the positive charges in the nucleus, so that atoms tend to have no electrical charge—they are neutral. Atoms normally cannot be divided by chemical means into simpler substances.

In natural radioactive elements such as the heavy metals radium and uranium, the nucleus has more energy than it needs, which makes it unstable. Nature's way of putting things in order is for part of the nucleus to break away spontaneously and release some of its excess energy. The process, called radioactive decay, continues until the nucleus is stable. For example, uranium and radium decay until lead is formed, which is stable, nonradioactive, and lighter in weight. The stable element at the end of radioactive decay is lighter because during decay some of the mass (weight) of the original nuclei is converted into energy, which is released as heat and other forms of radiation.

The amount of energy released is equal to the mass lost by the nucleus multiplied by the square of the speed of light. This is expressed in the famous equation of Albert Einstein, $E = mc^2$. In this equation, E is the symbol for energy, m for atomic mass, and c for the speed of

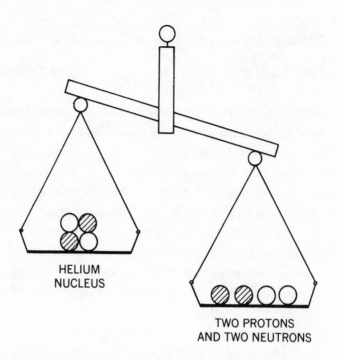

HELIUM
NUCLEUS

TWO PROTONS
AND TWO NEUTRONS

CONVERSION OF MASS TO ENERGY

*A case where the whole is not equal to the sum of its parts.
Two protons and two neutrons are distinctly heavier than a
helium nucleus, which also consists of two protons and two
neutrons. Energy makes up the difference.*

light. Because light travels at enormous speed (300 million meters per second), tremendous amounts of energy
are stored in the nucleus.

The energy of the nucleus (properly called nuclear,
rather than atomic, energy) can be released by two processes. One is nuclear fission, a special form of radioactive
decay. In a typical nuclear fission reaction, a stray neutron
from a radioactive nucleus hits another nucleus like a

bullet. The extra neutron makes the nucleus unstable and it splits into two "fission products" that fly apart. In addition, two or three new neutrons are released, which cause a chain reaction by hitting other nuclei and starting new fissions. The fissioning of uranium is the source of energy in nuclear power plants.

Nuclear fusion, in which two light nuclei are combined, or fused, to form the nucleus of a heavier atom, is the second process for releasing the energy of the nucleus. The first massive release of fusion energy was in the hydrogen bomb, exploded in 1951 by the United States. Since then, research to control the fusion process so that it can be used to provide power has been underway in the United States, as well as in many other countries. Developing fusion power plants may be the most difficult challenge that engineers and scientists have ever faced. If they succeed, it will not be for many years to come. Thus it seems likely that, until at least the year 2000, nuclear fission will be the only way of harnessing the almost unlimited energy locked in the nucleus of the atom.

Types of Radiation

Radiation is the emission of rays, wave motion, or particles from a source. Examples of radiation are light rays, x rays, radiant heat (such as that from radiators in a home), and particles smaller than atoms emitted by radioactive materials. The radiation released by radioactive materials can be in the form of neutrons and one or more forms identified by Greek letters alpha, beta, and gamma:

A CHAIN REACTION

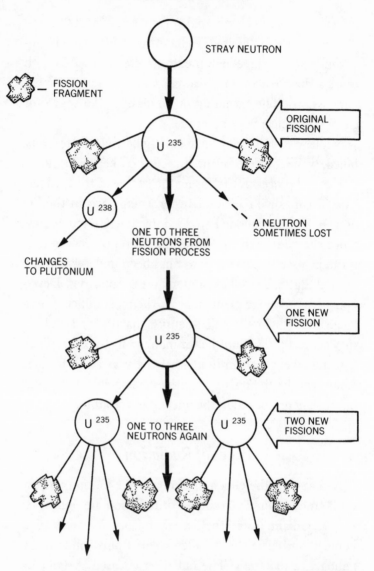

In a typical nuclear fission reaction, a stray neutron from a uranium 235 nucleus hits another uranium 235 nucleus like a bullet. The extra neutron makes the nucleus unstable, and it splits into two fission fragments that fly apart. In addition, one to three new neutrons are released, which causes a chain reaction by hitting other nuclei and starting new fissions.

16

- Alpha rays are streams of positively charged particles consisting of two neutrons and two protons bound together and hence are identical to helium nuclei. Alpha particles have the most energy of the three forms. They are also the largest and heaviest. Therefore they are more likely to collide with and be stopped by molecules of matter that they penetrate. Alpha rays can just penetrate the surface of the skin and can be stopped by a sheet of paper. Alpha particles are dangerous to humans, other animals, and plants only when the radioactive material releasing them enters a living organism. If the material enters the body by inhalation or along with food or water, the alpha particles released may damage the body.
- Beta rays are streams of electrons. Beta rays can burn the skin on contact and damage the body if the beta-releasing material is inhaled or swallowed. Most fission products release beta particles. A sheet of aluminum a few millimeters thick can stop beta particles.
- Gamma rays are similar to x rays, a form of radiation usually produced by bombarding a metal target with streams of electrons. X rays however, have less energy than gamma rays. Because of their high energy, gamma rays easily penetrate most materials. They can pass right through the human body but can be almost completely absorbed by 1 meter of concrete or other dense materials such as lead.

When these three forms of radiation—alpha, beta, and gamma—strike an atom or molecule, they can strip elec-

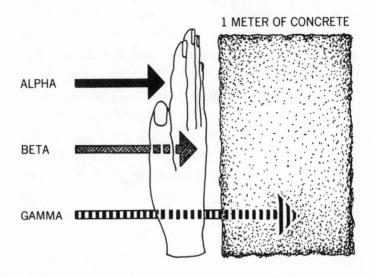

trons from them, producing positive ions. Ions are atoms or groups of atoms that have lost (or gained) one or more electrons and hence carry an electrical charge. Because of this, the three forms (and x rays as well) are called ionizing radiation. Other forms of radiation—for example, visible light, ultraviolet radiation, radio waves, and microwaves—produce less energy than ionizing radiation and so are less likely to damage the molecules that make up living cells, the fundamental unit of life. Stripping electrons from a cell's molecules interferes with the cell's chemical processes to the point, in extreme cases, of killing the cell. If enough cells die, the organism itself dies.

Biological Effects of Radiation

Ionizing radiation affects molecules in living cells in different ways. Because alpha particles are heavy, they

1 8

travel only short distances but can damage large numbers of cells in their path. Gamma rays, on the other hand, travel farther through the cell structure but are less likely to cause damage.

The kind of ionizing radiation is not the only factor influencing the harm that can be done. The amount is also important, including the rate at which the body receives it. In part, the amount depends on the half-life of the radioactive element, which is the time needed for one-half of its nuclei to decay to produce another element or a different physical form of the same element. For example, if one starts at noon with 1 gram of a radioactive element whose half-life is 10 minutes, 0.5 gram would remain at 12:10, 0.25 gram at 12:20, and only 0.125 gram at 12:30. At about 2:30 A.M., just a single radioactive atom would remain. Half-lives of different radioactive elements vary from a millionth of a second to more than a million years. Elements with short half-lives release a lot of radiation in a short time, but they also form stable elements very quickly. Those with long half-lives give off radiation much less rapidly because of the slow decay of their nuclei.

The effects of ionizing radiation on humans are commonly measured in rems, units that take into account how damaging the type of radiation is. Because a rem is a large quantity, millirems are also used; a millirem is one-thousandth of a rem.

Ionizing radiation can cause death immediately if exposure to the whole body (distributed relatively evenly) is more than 500 rems in a few minutes or hours. Exposures of this kind occurred when the two atomic bombs were dropped over Japan during World War II. If the

URANIUM 238 DECAY SERIES

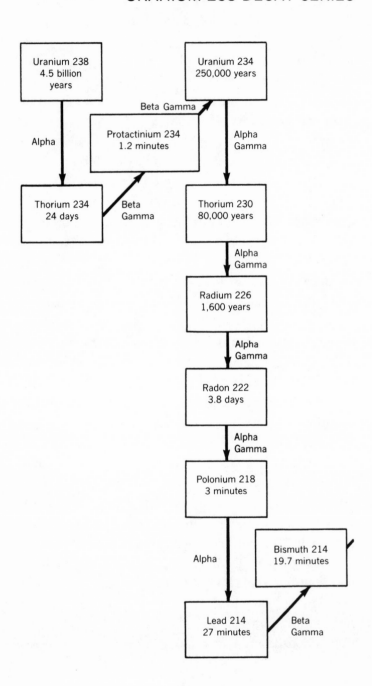

Most of the uranium in uranium ore is uranium 238, a radioactive isotope that decays over billions of years to become lead 206, a stable, nonradioactive element. The lengthy decay process passes through a series of intermediate elements called decay products, such as radium 226 and radon 222; these, too, are radioactive. The decay of uranium since the ore was formed millions of years ago has built up an inventory of the decay products identified here. In various concentrations, all are present in uranium mill tailings.

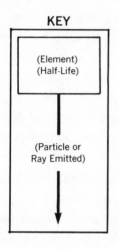

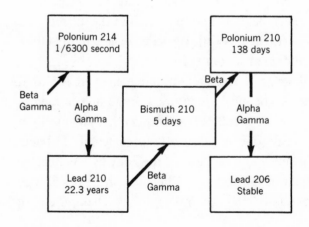

radiation exposure is between 100 and 500 rems, radiation sickness occurs. Typical symptoms are nausea, vomiting, diarrhea, changes in blood cells and, in later stages, susceptibility to infections, bleeding, and loss of hair. Radiation sickness frequently leads to death, but a slow recovery is possible.

The effects of lower levels of radiation are difficult to predict. The dose may kill a cell, which the body can then replace. Usually, the cell is only injured. In some cases, however, the injury leads to a mutation, a chemical change in the gene, the part of the cell containing the instructions that it needs to function and divide. When that injured cell divides, the mutation is passed on to the new cells that it forms. Even in adults who are no longer growing, most cells in the body must divide throughout life. Hair and fingernails grow constantly, and cells in the skin and lining of the mouth, throat, stomach, and intestines are constantly replaced.

Cancer (especially of the lungs, breast, and thyroid gland) and leukemia (sometimes referred to as "cancer of the blood") can be the consequences of faulty division of body cells. If the reproductive cells are injured, the mutation can be passed on to all cells in the offspring. So far, however, there is little evidence of hereditary effects in the children of A-bomb survivors.

The damage done by a radioactive element is also affected by its chemical properties. For example, a radioactive form of the metal strontium is similar chemically to calcium and so can substitute for calcium in bones. When it does so, it remains in the body as a radioactive constituent of the skeleton. Chemical properties also determine if a material can enter the food chain, the pathways by which that material, once it is absorbed by one plant or animal, travels through other plants and animals to humans. A radioactive form of iodine, for instance, if it is released in the air, easily passes through the food chain. It collects on grass, is eaten by cows, enters the human body in cow's milk, and then is concentrated in the thyroid gland.

Nuclear Fuel Cycle

Radiation is present throughout the nuclear fuel cycle, from mining and milling the ore through to the disposal of power plant wastes. Radiation levels are high, how-

THE NUCLEAR FUEL CYCLE

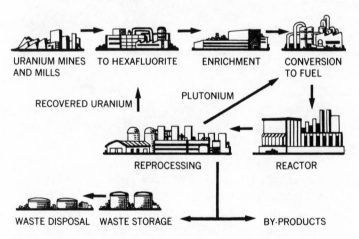

URANIUM MINES   TO HEXAFLUORITE   ENRICHMENT   CONVERSION
AND MILLS   TO FUEL

RECOVERED URANIUM   PLUTONIUM

REPROCESSING   REACTOR

WASTE DISPOSAL   WASTE STORAGE   ←   →   BY-PRODUCTS

Radiation is present throughout the nuclear fuel cycle, from mining and milling the ore to disposal of wastes. Radiation levels are high, however, only in the later stages of the cycle.

ever, only near the end of the cycle after the fuel has been fissioned in a nuclear power plant.

Uranium 235 is the fuel used in nuclear power plants in the United States. Like most elements found in nature, uranium is a mixture of different physical forms. The forms—called isotopes—differ in the number of neutrons (but not protons) in their nuclei and hence in their atomic weight. The number in the name of an isotope (235 in the case of uranium) is its atomic weight and in-

Uranium 235 is the fuel used by nuclear power plants in the United States. Here, at an open-pit mine near Grants, New Mexico, uranium ore is loaded into a dump truck. Later, the ore is ground and treated with chemicals to remove some impurities, usually at a mill near the mine. ERDA PHOTO BY WESTCOTT

dicates that there is a total of 235 protons and neutrons in its nucleus. Only 0.7 percent of uranium, the heaviest element found in nature, is uranium 235. Most of the rest is uranium 238, a slightly heavier isotope that does not fission.

After uranium ore is mined, it is ground and treated with chemicals to remove some impurities, usually in a mill near the mine. The uranium leaves the mill as uranium oxide (U_3O_8) and is called yellowcake because of its color and form.

Yellowcake is purified further and converted, by treatment with chemicals at high temperatures, to uranium

At a plant in Paducah, Kentucky, "yellowcake" is converted to uranium hexafluoride. Here, a worker adjusts equipment that produces the fluorine needed in the conversion, which is an important step in the production of fuel for nuclear power plants. OAK RIDGE OPERATIONS OFFICE—ATOMIC INDUSTRIAL FORUM

hexafluoride (UF^6), a material that changes from a solid directly to a gas at a low temperature (58° Celsius). The hexafluoride is transported to a third location where it is heated. When the gas is pumped through special filters, the uranium 235, which is lighter, passes through a little faster than the heavier uranium 238. After a while, gas that has gone through the first filter is richer in fissionable uranium 235. This gas is then pumped through a second filter. By repeatedly pumping the richer gas through a series of filters, the amount of uranium 235 is increased from the original 0.7 percent to 2 to 3 percent. This "enrichment" process requires thousands of pumps, and the gas must travel through thousands of kilometers of pipes. Some reactors use natural uranium as fuel, but this requires that the reactor be larger in size.

The enriched fuel is then transported to still another location, where it is converted by chemical treatment to uranium dioxide (UO^2). This material is formed into cylindrical pellets about the size of a pencil eraser and heated to produce a substance resembling a ceramic. The pellets are loaded end to end and sealed in long tubes about 1.3 centimeters in diameter and 3.7 meters long. The tubes are made of a zirconium alloy, which is a poor absorber of neutrons and so does not quench the fission reaction. The tubes are bundled together in a square pattern, close to each other but not touching, to form a "fuel assembly." Reactor Number 2 at Three Mile Island, which is typical of the newer reactors, held 102 metric tons of uranium contained in 177 fuel assemblies, each assembly holding 208 fuel rods. Up to this point, the fuel is not highly radioactive and can be handled relatively safely.

Increasing the amount of fissionable uranium 235 to 2 to 3 percent (in natural ores it is only 0.7 percent) requires very large plants such as this one in Oak Ridge, Tennessee. The "enriched" fuel produced by this process is used in commercial nuclear power reactors around the world. DOE PHOTO BY FRANK HOFFMAN

The fuel now is ready for use in a power plant, where it becomes part of the nuclear reactor. Except for the reactor, a nuclear power plant is similar to a conventional power plant that burns coal, oil, or natural gas to produce heat for use in making steam. In a conventional power plant, the heat for making steam comes from a chemical reaction, the burning of a fossil fuel; in a nuclear power

An engineer inspects fuel assemblies before they are inserted into a nuclear power reactor. A simple assembly holds about 200 rods, each loaded with pellets of enriched uranium fuel. SAMUEL A. MUSGRAVE, ATOMIC INDUSTRIAL FORUM

plant, the heat comes from a nuclear reaction, the fissioning of a mineral fuel. In both types of plants, the steam turns the blades of a turbine (toy pinwheels and windmills are simple kinds of turbines). The shaft of the turbine is linked directly to the shaft of a magnet in an electrical generator. As the steam spins the magnet inside coils of wire in the generator, electricity is continuously produced from the kinetic energy (the energy of motion) of the spinning magnet.

Levels of radiation rise dramatically inside the nuclear reactor. The two major fission products created when a uranium nucleus splits are radioactive and undergo one or more steps of radioactive decay before becoming lighter, stable, nonradioactive elements.

Not every uranium nucleus splits exactly the same way,

so a number of fission products are formed in a fuel pellet. Those with short half-lives may almost completely vanish within minutes or a few hours, leaving behind different elements that may or may not be radioactive themselves. Fission products with long half-lives are another matter. For example, strontium 90, a common fission product, has a half-life of 28 years. It takes 10 half-lives—280 years—for its radioactivity to drop to less than one-thousandth of the original amount. Eventually,

A reactor is being loaded with nuclear fuel. The reactor is in the bottom of the pit (foreground). The top of the reactor containing the control rods (background) is in its storage position. After the fuel is loaded, the top is replaced. ATOMIC INDUSTRIAL FORUM

however, all the radioactive elements decay to form stable elements that no longer release radiation.

A second type of element formed in the pellets is an even longer-term problem. Some uranium 238 nuclei (which are not fissionable) absorb the neutrons striking them, creating heavier elements not found in nature. The most important of these is plutonium, which is fissionable and can join in a chain reaction. Plutonium 239, the isotope that accounts for 60 to 70 percent of the plutonium created in nuclear fuel, has a half life of 24,400 years. If inhaled or swallowed, it can cause cancer.

Despite the creation of fissionable plutonium in the fuel pellets, the chain reaction begins to slow down; as the products created by the fission build up, they soak up extra neutrons without releasing any energy. In a typical nuclear reaction, up to one-third of the assemblies must be replaced each year to keep the fission reaction at the speed desired.

"Spent fuel" (as it is called) contains some of the original fissionable uranium, a similar amount of fissionable plutonium, other synthetic elements heavier than uranium, and lighter elements resulting from fission. The fuel is hot and highly radioactive.

To remove spent fuel, a compartment near the top of the reactor is filled with water, which cools the assemblies and also acts as a shield to prevent radiation from escaping into the environment. The top is removed and the fuel assemblies are transferred through a water-filled tunnel to a nearby room. There they are lowered into a deep "swimming pool" filled with water. The pool is made of reinforced concrete lined with stainless steel.

With the reactor top removed, used fuel can be withdrawn
from the reactor core and replaced with assemblies of uranium
dioxide fuel. A typical power reactor undergoes one such
refueling a year, at which time up to one-third of the fuel is
replaced. ATOMIC INDUSTRIAL FORUM

An engineer positions spent fuel assemblies in the "swimming pool" of a nuclear power plant. The water in the pool cools the assemblies and also acts as a shield to prevent radiation from escaping into the environment. ATOMIC INDUSTRIAL FORUM

For a few days, the pool glows with a pale blue light from Cerenkov radiation, named for the Soviet physicist who explained the glow produced by some charged particles as they travel through water.

The fuel decays rapidly, and the glow disappears in a few days. After 6 months or so, about 99.9 percent of the original radiation has vanished through decay. That remaining, however, is dangerous for a thousand years or more.

Once the spent fuel assemblies have cooled and their radiation decreased substantially, they can be transferred to special casks and shipped to a chemical reprocessing plant. There the rods can be cut into short pieces, and the exposed fuel pellets dissolved in nitric acid. Almost all of the uranium 235 and plutonium could be recovered for reuse as nuclear fuel. Some spent fuel produced by foreign countries and by the United States Government in its nuclear weapons program has been reprocessed. But almost all spent fuel from United States nuclear power plants is still stored in pools at the plants themselves because no reprocessing plants are in operation in the United States.

Decommissioning

A nuclear power plant (as well as a nonnuclear power plant) is expected to operate for 30 to 40 years. During that time, neutrons bombard not only the fuel but also other parts of the reactor and reactor building, making them radioactive. When the time comes to shut down the plant, it must be done in a way that protects the environment from these radioactive materials, which can remain hazardous for long periods of time. The process of shutting down a plant, called "decommissioning," can be done in several ways. First, radioactive fuel and wastes are removed. Then the plant can be "mothballed" by locking and guarding it to prevent anyone entering. Or the plant can be "entombed" by covering it over and sealing it with concrete. Or it can be "dismantled" and carted away for disposal elsewhere. Dismantling may be delayed by mothballing or entombing the plant for 30 to 50 years to allow the shorter-lived radioactive elements to decay before the wastes are sent elsewhere for disposal.

Dismantling a reactor involves both conventional and specialized methods. Conventional demolition and salvage techniques can be used for equipment and buildings that are not radioactive. Scrap steel, equipment, and construction material can be sold or reused if it is not radioactive, just as it is for any industrial plant. The core and other radioactive parts must be dismantled, using remote control equipment. The radioactive material on the surface is removed by cleaning with chemicals, scraping, jets of high-pressure water, and other techniques. Special measures are taken to prevent the spread of radioactive dust and gases. For example, some of the components are submerged in water while being dismantled with a special, remotely controlled torch.

Until recently, it was assumed that the cheapest way to get rid of a worn-out reactor was to entomb it. Most of the highly radioactive materials in the reactor were thought to be due to the cobalt that is present in most steels. Cobalt 60, which is produced when steel is hit by neutrons, has a half-life of 5.3 years. Thus, within a few decades, radiation would have dropped to safe levels.

Nuclear engineers knew that other radioactive isotopes would be formed but thought that they would be formed in such tiny amounts that they would contribute very little radioactivity. Then in the late 1970s, two other potentially important isotopes were identified: nickel 59, with a half-life of about 80,000 years, and niobium 94, with a half-life of 20,000 years. They are indeed present in tiny amounts, but their long half-lives mean that they could be giving off significant amounts of radiation long after cobalt 60 has decayed to safe levels. Entombing would be acceptable, therefore, only if the long-lived isotopes could be removed or if the "tomb" could last thousands of years, which is highly unlikely. Although it is cheaper to delay dismantling for several decades until radiation has dropped, the savings are lost by the cost of guarding and maintaining the tomb all those years.

Since 1960, about 70 nuclear reactors have been decommissioned in a number of countries. No large commercial reactor has yet been decommissioned in the United States.

In the next few years, the Federal Government plans to dismantle the commercial nuclear power reactor in Shippingport, Pennsylvania. When it began operating in 1957, it was the first commercial nuclear power plant in the United States. The dismantling is expected to take about 5 years to complete, to cost more than $40 million at today's prices, and to produce almost 12,000 cubic meters of radioactive waste. This is as much waste as will be produced by the cleanup of the damaged reactor at Three Mile Island, which generates about 75 times more power than Shippingport does. Because it is similar in design to modern reactors and has been in operation for many years, the dismantling of the Shippingport reactor will provide valuable experience in the problems of tearing down other large worn-out reactors at the end of their operating lives.

III.

Accidents:
The Now Problem

THE POSSIBILITY of a major accident in a nuclear reactor has been a concern since the beginning of the nuclear age. Such an accident would not be like the explosion of an atomic bomb. In a simple atomic bomb, two or more pieces of almost pure fissionable material, usually plutonium, are rapidly brought together and held together for the short time needed to generate a very large explosive force. In a nuclear power reactor, however, the fuel contains only 2 to 3 percent of fissionable material, and there is no way of bringing and holding the pieces together.

How a Reactor Works

The purpose of a nuclear power reactor is simple: it provides a setting in which a chain reaction can be started, sustained, and controlled, and from which the heat produced can be recovered to generate steam. The essential components of a reactor are a "core" of fuel, a moderator, a control system, and a system for removing heat, all contained in a reactor vessel that is surrounded by shielding and containment structures to prevent radiation from escaping.

The *fuel* is the heart of the reactor. In a typical large reactor, the core consists of about 200 fuel assemblies. Because the tubes in each assembly are close together, they can capture one another's neutrons to keep the chain reaction going. The neutrons in a chain reaction are released at very high speeds. They lose speed as they collide with the surrounding matter in the reactor core. This loss of speed is desirable because slow-moving neutrons are better at triggering the chain reaction than fast-moving neutrons. If there are too many collisions, however, a neutron may bump into a nucleus and be absorbed by it without producing energy. Thus, a reactor needs a *moderator*, a material that can slow down neutrons quickly but absorbs few of them. Ordinary water circulating around the fuel tubes serves as the moderator in almost all United States nuclear power reactors and so they are called light water reactors. Another type of reactor, developed largely in Canada, uses water containing extra amounts of a heavy isotope of hydrogen and is known as a heavy water reactor.

The *control system* of the nuclear reactor consists of rods of materials that are good absorbers of neutrons. By absorbing the moving neutrons, the rods can slow or stop the fission reaction. Boron and cadmium are often used in control rods. They are moved into and out of the core as necessary to control the amount of heat produced

PRESSURIZED WATER REACTOR

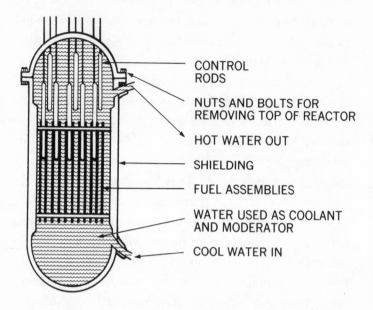

CONTROL
RODS

NUTS AND BOLTS FOR
REMOVING TOP OF REACTOR

HOT WATER OUT

SHIELDING

FUEL ASSEMBLIES

WATER USED AS COOLANT
AND MODERATOR

COOL WATER IN

by the chain reaction. Typically, a reactor is equipped
with regulating rods that control the chain reaction routinely and extra safety rods that shut the reactor down
quickly and completely in an emergency.

Most of the energy released in a chain reaction is generated by the fission products moving at a high speed in
the fuel. As the fission products collide with surrounding
matter, this kinetic energy is almost instantly converted
to heat energy. The *heat removal (or coolant) system*
uses the same water that serves to moderate the chain reaction. By circulating around the tubes, the water picks
up heat and prevents the fuel from melting, which would
increase the possibility of its leaking out of the reactor.

The heat removed from the core is used to boil water to steam in steam generators.

A steel *reactor vessel* holds the core, control rods, and circulating water. The vessel of the Three Mile Island reactor is 12 meters high, with walls 22 centimeters thick.

Two giant steam generators separated by an equally massive reactor vessel are being barged to the construction site of a nuclear power plant. This shipment, weighing about 950 metric tons, traveled about 2,100 kilometers between Chattanooga, Tennessee, and the Fort Calhoun plant in Nebraska, which started operations in 1973. ATOMIC INDUSTRIAL FORUM

It is surrounded by two separate concrete-and-steel shields, with a total thickness of almost 3 meters. The *shielding* prevents the escape of penetrating radiation emitted primarily by fission products. The reactor vessel and steam-generating system at Three Mile Island are set inside a *containment building*, almost 60 meters high

A typical view of a nuclear power plant under construction shows some of the many miles of steel reinforcing bars that are embedded in concrete to add strength to the structure. The reactor building of the plant—the Diablo Canyon station in northern California—looms in the background. ATOMIC INDUSTRIAL FORUM

and made of reinforced-concrete walls 1.2 meters thick. The building is the final barrier that prevents radioactive materials from escaping to the environment.

Defense in Depth

In the United States, nuclear power has been largely developed, promoted, and regulated by the Federal Government. Few other activities, if any, are as tightly controlled by the Government. A company that wants to build a nuclear power plant must receive permits from the United States Nuclear Regulatory Commission (NRC), as well as from state and local governments. When the plant is completed, it cannot start operating until NRC issues an operating license. Plants are inspected regularly by NRC while being built and after they begin operations. Most people agree that the Government's safety standards are the most severe required of any industry in the world. These standards may be even more severe in the future.

Nuclear power plants are designed and operated according to a principle called "defense in depth," the same principle that guides the United States space program. The design of a plant assumes that its equipment can fail, that people will make mistakes in operating it, and that severe natural events—earthquakes and floods, for instance—can damage it. Instruments built into the plant record temperature, pressure, water level, and other factors that affect safety. These instruments are linked to control systems that can immediately and automatically adjust or shut down the reaction if its operations are not within the normal range. If any part of the system fails, back-up systems take over to prevent accidents or to lessen the effects should an accident occur.

United States nuclear power plants have been built to withstand the loss of cooling water, which is the most severe accident that engineers can imagine. In such an accident, one of the large main pipes feeding cooling water to the reactor would be broken in two and the broken ends would swing apart. Water pouring from both ends would produce clouds of radioactive steam. The reactor would automatically stop at once, but its fission products would continue to decay and produce heat. If the emergency cooling system then failed to flood the reactor core with water, as it is designed to do, the temperature in the core might soar to several thousand degrees. Depending on how long it remained uncovered, this could melt the radioactive core. It would then be possible for the mass of molten metal to burn its way through the many protective "envelopes" surrounding the reactor. Radioactive materials then could be released, causing serious harm to all living organisms they encountered.

No such "meltdown" of a nuclear reactor has ever occurred anywhere in the world. In fact, defense in depth has worked so well that nobody inside or outside a commercial nuclear power plant, as far as is known at present, has ever been injured by radiation.

After the accident at Three Mile Island, nevertheless, both critics and supporters of nuclear energy were forced to take a new hard look at the safety of present nuclear plants and the design of future plants.

Three Mile Island

Three Mile Island sits in the Susquehanna River about 16 kilometers from Harrisburg, Pennsylvania. Since early in this century, it has been owned by power companies. Until a small hydroelectric dam was built across the Susquehanna to the Island, it could be reached only by boat.

When the region needed more electricity, General Public Utilities Corporation decided to build a nuclear power plant on the Island. One of its member companies, Metropolitan Edison Company ("Met Ed"), began operating one unit—known as "TMI-1"—in 1974, and another, TMI-2, in late 1978. Together, the two can generate 1,700 megawatts, about enough electricity to serve a city of 2 million people. Some critics of nuclear power claim that TMI-2 was hurried into operation, and in the early months of 1979, TMI-2 did experience a number of problems. By late March 1979, however, it was in full operation.

A large nuclear power plant can be described as a "plumber's nightmare." Each TMI unit is as big as a medium-sized shopping center. Inside is a maze of pipes as big around as manhole covers and valves as big as fence gates to regulate the flow of gases and liquids through the pipes.

The TMI reactor contains two separate water loops that meet in each of two steam generators. Water in the reactor coolant system—the "primary loop"—flows around each fuel rod in the reactor core and then through a 90-centimeter diameter steel pipe to the steam generators nearby. There it flows through a series of small tubes.

43

Nuclear power reactors, like TMI-2, are sometimes called "plumbers' nightmares" because of the maze of pipes they contain. DEPARTMENT OF ENERGY

Water in the other loop—the "secondary" or "feedwater" loop—flows around these tubes containing hot water from the primary loop, picks up some of its heat, and boils. Thus water in the secondary loop is converted to steam without ever directly contacting the radioactive water in the primary loop.

This nonradioactive steam leaves the two steam generators and travels out of the containment building into the turbine building, where it enters the turbine, which turns the electrical generator. After the steam has passed through the turbine, it enters a condenser. There it is cooled and becomes water again and is then recycled back through the steam generators. The condenser is cooled by water from yet another closed loop, the tertiary loop. Having been warmed in the process of condensing steam to water in the feedwater loop, the water in the tertiary loop is pumped up to the cooling towers, the most visible feature of many power plants, both nuclear and nonnuclear. In the cooling tower, the tertiary water tumbles down a series of steps, giving off clouds of water vapor that escape out of the top of the cooling tower. Water from the cooling towers is pumped back to the condenser, where it is recycled.

Meanwhile, the primary coolant water, now at a temperature of about 290° C, flows back to the reactor core, where it picks up heat from the fission process and rises to a temperature of about 318° C. Then the cycle is repeated. Although the water is hotter than its normal boiling temperature (100° C at sea level), it does not boil because it is kept under pressures about 150 times higher than normal.

Water at high pressure can be heated far beyond normal boiling temperature, and this permits the plant to produce steam at higher temperatures than is possible were the primary cooling water allowed to boil at 100° C. This is similar to the way a pressure cooker can reach higher temperatures than can water boiled in a pan.

"Pressurized water reactors" of this type are used in about 60 percent of United States nuclear power plants. Most of the rest are boiling water reactors, in which water circulating through the reactor boils and the resulting steam spins the turbine-generator. The two reactors are different kinds of light water reactors.

The pressure in the TMI reactors is maintained by a

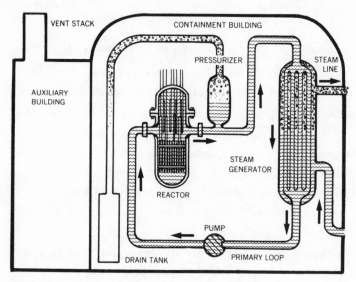

POWER PLANT NUMBER TWO AT THREE MILE ISLAND

pressurizer, a large tank located between the reactor and
the steam generators. The pressurizer is half full of coolant water. Above the water is a bubble of steam. Heating
the water slightly makes the bubble larger and so increases the pressure in the primary coolant loop. Cooling
the water slightly makes the bubble smaller and decreases
the pressure.

The accident at TMI-2 began on Wednesday, March
28, 1979, at 4:00 A.M. In the turbine building, crews were
doing maintenance work on equipment in the feedwater
system. Suddenly, there were sounds described as "loud
thunderous noises, like a couple of freight trains." Later,
the noise was traced to "water hammer," a familiar sound
in any house where air has gotten into hot water pipes.
Water hammer can occur in large pumping systems when
the flow of water is suddenly stopped—as just had hap-

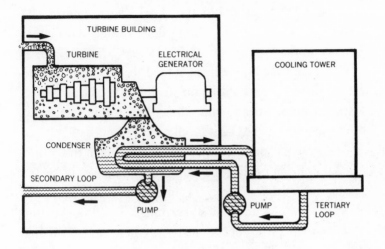

*Power Plant Number 2 at Three Mile Island contains two
separate water loops that meet in each of two steam generators
(only one shown). Water in the primary loop flows around
each fuel rod in the reactor core, is heated and then moves to
the steam generators. There it flows through a series of small
tubes (not shown). Water in the secondary (or feedwater)
loop picks up some of the heat in the primary loop and boils.
Thus water in the secondary loop is converted to steam
without ever directly contacting the radioactive water in the
primary loop. This nonradioactive steam leaves the two steam
generators and travels to the turbine building, where it enters
the turbine, which turns the electric generator. After the
steam has passed through the turbine, it enters a condenser,
which is cooled by water in the tertiary loop. The steam is
cooled in the condensers and becomes water again, then is
recycled back through the steam generators.*

pened in the feedwater loop of TMI-2 when pumps were stopped because of something done during the maintenance work.

With no feedwater being added to the steam generators, there soon would be no steam, so the plant's safety system automatically shut down the turbine-generator. The incident at TMI-2 was only a few seconds old.

Meanwhile, other events were happening in the control room. The horseshoe-shaped control panel is 12 meters long and consists of dials, gauges, and 1,200 lights and alarms that sound or flash warnings. Reactor operators are trained to respond quickly to emergencies. On that morning in March, 1979, about 100 alarms sounded within minutes.

With no flow of feedwater to take away the heat, the water in the primary loop, which is supposed to cool the core, became hotter. The rapidly heated water expanded, and the level of water inside the pressurizer rose, compressing the steam above it. As pressure built up, a valve atop the pressurizer opened—as it was designed to do—to relieve the pressure. First steam and later water began flowing out of the reactor coolant system through the relief valve and down a drain pipe to a tank on the floor of the containment building. Pressure continued to rise, however, and 8 seconds after the feedwater pumps stopped, TMI-2's reactor "scrammed": its control rods automatically dropped down into the reactor core to halt the chain reaction. Less than a second later, practically no heat was being generated by the reaction itself. As in any nuclear reactor, however, the fission products continued to give off heat and raise the temperature of the

CLOSE CALL

Three Mile Island was not the first serious accident at a nuclear reactor. One earlier accident in the United States caused three deaths, and accidents here and in other countries have resulted in damage to the core, measurable releases of radiation, or contamination after the accident at least as great as at TMI-2. Most of the accidents involved reactors that were small, located in remote areas, and government-owned or financed.

National Reactor Testing Station, Idaho Falls, Idaho

1961: The first major nuclear reactor accident in the United States occurred at an experimental military reactor at the National Reactor Testing Station in Idaho Falls, Idaho. While the reactor was shut down to repair the control system, servicemen apparently made a mistake that resulted in the control rods being lifted out of the core. The fuel melted and there was a steam explosion. Two workers were killed in the building and a third died on the way to the hospital. The deaths resulted from physical injuries, although the radiation levels also would have been fatal. One body was not recovered until the sixth day after the accident. During the explosion, the worker had been pinned to the upper part of the reactor building, directly above the reactor. The body was removed by eight men, paired in quick-moving relays to avoid excessive exposure to radiation. No two-man team was in the building more than 65 seconds. The radiation was contained on the site, so there was no public danger. Decontaminating the site took 19 months.

reactor's coolant water. This heat amounted to only 6 percent of that released during fission, but it still had to be removed to keep the core from overheating.

With the reactor scrammed and the relief valve on the pressurizer open, pressure in the reactor coolant system fell. Up to this point, the reactor system was responding normally to the stopping of the turbine. But the relief valve should have closed 13 seconds into the accident, when pressure dropped to normal. It did not. A light on the control panel indicated that the electric power that opened the valve had gone off, which led the operators to assume that the valve had shut. But for some reason, it was stuck open and remained open for 2 hours and 22 minutes, draining precious coolant water from the reactor at the rate of 830 liters per minute and flooding the floor of the containment building. A dreaded loss-of-coolant accident was in progress. In the first 100 minutes of the accident, more than one-third of all the water in the reactor coolant system escaped through the valve.

With the valve stuck open, pressure continued to drop. As it did, TMI-2's automatic systems once again went into action. Two pumps, part of the emergency core-cooling system, started pushing new water into the primary coolant loop at the rate of 3,800 liters per minute to ensure that the reactor core remained covered. With the pumps working, the level of water in the pressurizer rose. The operators in the control room of TMI-2 had been trained never to let the pressurizer fill with water because then there would be no steam bubble in the pressurizer and they would lose the ability to regulate pressure in the system through the control of the bubble.

CLOSE CALL

Chalk River, Ontario

1952: An accident occurred at a research reactor at an Atomic Energy of Canada, Limited plant near Chalk River, Ontario. Experiments were being conducted at a low power with the flow of cooling water reduced. Through a misunderstanding, an operator suddenly pulled out the control rods, permitting power to build up rapidly in the reactor. Several fuel rods overheated and burst, sending fission products into the cooling water. Before order could be restored, about 3.8 million liters of radioactive water poured into the basement beneath the reactor. The volume was the same as at TMI-2, but the water was not as radioactive as that at TMI-2. The water was pumped to a disposal area on the site. The accident also sent large amounts of radioactive gases up the plant stack. Fortunately, unusual atmospheric conditions prevented the gases being dispersed over a wide area. The reactor was dismantled using unique tools designed for removing the damaged core. To reduce the radiation exposure to each individual, about 1,000 military men were called in to help in the cleanup. The reactor was back in operation 14 months later.

So about 2½ minutes after the pumps started working, the operators shut one down, reduced the flow in the second, and began draining the reactor's cooling water.

The training and written emergency procedures of the operators in the TMI-2 control room had never dealt with the possibility of an accident in which coolant was lost through the top of the pressurizer, so that the operators repeatedly misunderstood what was happening. Of course, the relief valve that failed to close as it was designed to do was crucial in what happened. But if the control room operators had realized that the valve was stuck open, they could have closed a backup valve to stop the flow of coolant water. Or if they had simply left the emergency pumps on, the accident at Three Mile Island would have remained little more than a minor inconvenience.

Within minutes, with the relief valve stuck open and pressure continuing to drop, the coolant began to boil. At about one hour into the accident, the four pumps that circulate water through the primary loop began to vibrate because they were pumping a mixture of steam and water produced by the boiling. These pumps are extraordinary devices. Each is about the size of a dump truck, costs several million dollars, and consumes enough electricity to light a small town. At about 5:15 A.M., 1¼ hours into the accident, the operators turned two of the pumps off to prevent damage from vibration; one half hour later they turned the last two off.

When the pumps were working, a mixture of steam and water flowed through the core, cooling it enough to keep it from melting. When the flow stopped, the steam separated from the water, with the steam rising to the upper part of the reactor vessel, like gas in a Coke bottle that has been shaken. Meanwhile, water continued to pour out of the stuck-open valve. The upper parts of the core were soon uncovered and bathed in rising steam. A chemical reaction took place between the very hot zirconium in the fuel rods and the steam. The oxygen in the steam combined with zirconium to form zirconium oxide, and hydrogen gas was released, which can explode if oxygen and a source of ignition are present. The gas tended to block the flow of cooling water around the superheated core and further delayed its cooling. The reaction weakened the fuel rods, which broke open and released radioactive material into the water. Eventually, parts of the core may have collapsed into a jumbled pile near the middle of the core, forming a sort of crust that may later have blocked the upward flow of coolant water. This can be verified only when the reactor is opened.

The first major correct move of the day came at 6:22 A.M. Exactly why is uncertain. One explanation is that a new supervisor coming into the control room at 6:00 A.M. concluded that the relief valve on the pressurizer was leaking and ordered the backup valve closed. Another explanation is that no one could think of anything else to do to bring the reactor back under control. In any event, the loss of coolant was stopped, and pressure in the reactor cooling system began to rise.

Shortly before 7:30 A.M., the emergency pumps were turned back on, and later, sometime during Wednesday, the first day of the accident, the core was again completely covered with water. By then the damage had been done, but not until late on Thursday or early on Friday did those handling the accident fully understand what had happened and the dangerous conditions in the core of the TMI-2 reactor.

Meanwhile, the operators still did not realize that the core had been uncovered, failing to understand evidence that should have told them otherwise. What they did

CLOSE CALL

Windscale, England

1957: A fire started at a reactor producing plutonium in Windscale, England, when some of the fuel rods overheated. The fire, which started when the reactor was shut down for maintenance, was caused by an operator's error and inadequate instruments. The accident resulted in the largest known releases of radioactive gases into the environment from a nuclear reactor. The amount of iodine 131 from the plant's stack was well over 1,000 times that estimated for TMI-2. The total dose of gamma radiation to persons in the region of heaviest contamination was about one-tenth the maximum permissible exposure. On the site, the average level of air contamination was about twice the daily standard.

The United Kingdom Atomic Energy Authority temporarily banned consumption of milk in an area of 500 square kilometers to avoid contamination from iodine 131. In contrast to what happened at TMI-2, good public relations were maintained throughout the incident, and the public remained confident in the authority.

A second plutonium-producing reactor at Windscale, which was not affected by the fire, was shut down while the accident was investigated. The authority concluded that it would be too expensive to change the second reactor to prevent a similar type of fire. It was permanently shut down. All the fuel from the second reactor and the undamaged fuel from the first reactor were removed, and both reactors were sealed with concrete.

know was that radioactive water was in an auxiliary building where it was not supposed to be. It got there when,
7½ minutes into the accident, pumps in the containment
building automatically started to move the water on the
floor over to tanks in the auxiliary building. Before the
pump was turned off, 1.5 to 1.9 million liters of slightly
radioactive water were pumped into the auxiliary building.

Soon, higher-than-normal levels of radiation were being
recorded throughout the plant. At 6:45 A.M., plant officials declared a "site emergency," as required by TMI's
emergency plan whenever some event threatens "an
uncontrolled release of radioactivity to the environment."
At 7:24 A.M., a general emergency was declared because
conditions had the "potential for serious radiological
consequences to the health and safety of the general
public." State and Federal officials were notified, and
many soon appeared at Three Mile Island, as did re-
porters—400 before the incident was over. The accident
at Three Mile Island became a major news story of 1979.

During the rest of the first day, the operators struggled
to bring the reactor under control. Finally, 16 hours after
the accident started, they succeeded in circulating water
through most of the core and the rest of the primary loop.
Everyone heaved a sigh of relief, thinking the emergency
was over. They still did not know that the core had suffered severe damage.

Thursday started off relatively calmly. Officials thought
that they had the reactor under control. Radiation levels
remained high in the auxiliary building. Also, some
slightly radioactive water had been discharged into the

Susquehanna River. Generally, however, radiation levels outside the plant indicated no problems. But by the end of the day, analysis of the reactor's coolant water showed that core damage was far greater than had been thought.

Friday, March 30, is a day most people who lived near Three Mile Island will never forget. Wednesday, the day of real danger, had had a lot of suspense, some fear, and many rumors. By the evening, however, local radio and TV stations were saying that the accident was over. Friday shattered Thursday's calm.

Throughout the episode, there were frequent cases of confusion, poor communications, and failure of those in charge to understand or correctly interpret the information they were receiving. Friday had more than its share of confusion and poor communications. Met Ed decided that some radioactive gases had to be released from the auxiliary building. Although the amounts were small, a series of mistakes and confusion led the Nuclear Regulatory Commission to recommend evacuating people living within 16 kilometers of Three Mile Island. When some of the confusion cleared, the governor of Pennsylvania recommended that pregnant women and preschool children (who are more sensitive to radiation) leave the area within 8 kilometers of the plant. He also recommended that schools be closed. Even these measures probably were not necessary.

Friday was also the day fear developed over a bubble of hydrogen gas that had collected in the reactor. No oxygen was present, and there was no source of ignition. Still, NRC feared it might explode. The explosiveness of hydrogen is well known—it was responsible for the burning of the *Hindenburg*, a German zeppelin, in New Jersey in 1937. Soon the hydrogen bubble was making headlines and increasing public fear. It is quite possible that the most serious health effect of the Three Mile Island accident was and will be the severe mental stress suffered by the people living nearby. Adding to the stress, by coincidence, was *The China Syndrome*, a popular movie at the time. The movie dealt with a nuclear reactor in which a meltdown, supposed to be capable of wiping out "an area the size of Pennsylvania," was narrowly averted. In such a meltdown, a fiercely hot molten core would eat its way through all the barriers and end up in China at the opposite side of the globe.

On Sunday, April Fool's Day, after more confusion and contradictory statements from various officials, everyone agreed that there was no risk of a hydrogen explosion. At the same time, the hydrogen bubble appeared to be getting smaller in size. The hydrogen still existed but was scattered throughout the system in smaller bubbles that were easier to eliminate.

By Monday, April 2, the crisis was over, but the threat to the health and safety of the workers and citizens did not entirely disappear. A small bubble of hydrogen remained and the reactor was badly damaged. The reactor had to be brought to "cold shutdown"—the point at which the temperature of the coolant falls below its boiling point. When that took place a few days later, the problem did not end. More than 3.8 million liters of radioactive water remained inside the containment building or stored in other buildings. The containment building also held radioactive gases and the badly damaged and highly radioactive core. The cleanup, unlike anything in the history of the nuclear power industry, is estimated to cost at least $1 billion and to take as long as 10 years.

Cleanup at TMI

Much of the technology to be used in the cleanup at Three Mile Island is based on methods developed principally at facilities owned by the Federal Government for other applications, including previous accidents. The technical task is in many ways the most manageable part of bringing the accident to TMI-2 to an end, but it is complicated by financial, social, legal, and regulatory factors that pose many conflicts and have few clear answers.

The damaged reactor represents a potential danger, most directly to the cleanup crews but possibly to the public as well. Thus, it is desirable for cleanup to proceed as quickly as possible. On the other hand, the job is large and complex, involving many controversial issues, and it is taking place in an atmosphere of public anxiety and distrust. Therefore, it also is necessary for the owners of TMI and NRC, which must approve all cleanup measures, to be cautious, plan carefully, and provide an opportunity for the public to express its opinions.

Cleanup started as soon as the reactor was stabilized. The slightly radioactive water in the auxiliary building was decontaminated using equipment that works much like a home water softener. The equipment contains a series of filters that trap radioactive materials on a bed of resin particles. The buildings, except for the containment building, were cleaned up, mostly by dry and wet vacuuming, mopping, and wiping of radioactive areas by workers wearing protective clothing and breathing equipment. The clothing, rags, resins, and other materials the workers used were trucked to Richland, Washington, where a private company operates a burial ground for materials with low levels of radioactivity. Two other burial grounds—in Barnwell, South Carolina, and Beatty, Nevada—were in operation, but the governors requested that no wastes from the cleanup be sent to their states.

The containment building was a far more difficult

Workers had to wear protective clothing in the clean-up of the damaged reactor at Three Mile Island. DEPARTMENT OF ENERGY

problem. At first, entering it with robots was considered. But later, plant officials concluded that the radiation levels were low enough to permit workers inside for brief periods. First, the radioactive gases had to be removed. After considering several alternatives, NRC approved venting the gases to the atmosphere through the plant's stack. Venting aroused considerable controversy, but NRC insisted it could be done with no damage to the surrounding area because much of the radioactivity had disappeared by decay.

Next, the 2.6 million liters of highly radioactive water in the containment building had to be removed and decontaminated by pumping it first through one bed of resin, then through a second. Processing of the water started in September, 1981, and was completed in mid-1982. The resins become highly radioactive and cannot be stored permanently on an island in the middle of a river in a densely populated area. In addition, the geology of Three Mile Island is not suitable for burial of the wastes. The Federal Government agreed to accept the wastes at its Hanford Reservation, near Richland, Washington. The resins will be used in studies of methods for treating wastes to make them suitable for burial.

The cleanup plan calls for the first workers inside the containment building to wash down the surfaces with water or steam. The workers will wear three layers of protective clothing and breathing masks. Then other workers will try to reduce radiation levels still further and clean up "hot spots" with liquids containing detergents, solvents, or other chemicals. The workers often will have to shield themselves from one hot spot while trying to

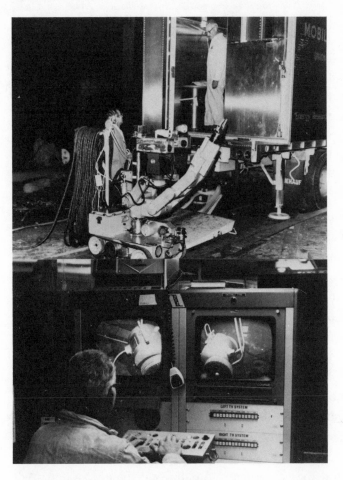

A *mechanical remote-controlled industrial robot known as*
"Herman" can be used in environments with high levels of
radiation. The manipulator is designed to operate at distances
up to about 210 meters from its control console (lower
photo), to which it is attached by a cable. It has a mechanical
hand capable of lifting 72 kilograms and dragging up to 230
kilograms. Motions with the hand include shoulder, wrist
and hand movement, directed by controls on the console.
Two television cameras are mounted behind the arm to
transmit pictures to monitors on the console. DOE PHOTO
BY FRANK HOFFMAN

clean up another. For shielding, they will use lead bricks, thick rubber mats, steel plates, or concrete blocks. At the peak of the cleanup, as many as 2,500 workers may be needed. They will be continuously rotated to keep their exposure to radiation well below the maximum permissible dose.

The cleanup may require huge quantities of materials— for example, 1 million pairs of plastic coveralls, plastic boots, and rubber gloves, plus thousands of breathing masks. As much as 1 million liters of special cleaning liquids may be needed. The liquids will pose a special problem because they contain chemicals that make decontamination by the use of resin beds ineffective. Thus another complex piece of equipment will have to be constructed to evaporate the liquids to a solid material that can be transported to a Federal burial ground.

The greatest uncertainties will occur when the top of the reactor vessel and the badly damaged core have to be removed. Great care will have to be taken to guard against starting a chain reaction. The actual steps will be decided after the reactor is inspected, which is impossible until the containment building can be decontaminated. Workers must first disconnect components such as the control rods that run through the top of the vessel into the core. These components may be jammed or entangled if the core has been distorted.

All work on the core will be performed remotely under water with the aid of TV. The radioactive materials will be encased in metal containers underwater and their ends welded shut with underwater welding equipment. The cans will be placed in special shipping casks and transported to a Federal burial ground.

Removal of the core is the last step in cleanup. Only then can officials of the plant assess the damage to TMI-2 and decide its future. They hope to put TMI-2 back in operation using the undamaged portions of the plant. Over $700 million had been invested in TMI-2, and it had been in operation only for a few months at the time of the accident. If the plant can be used, the existing reactor can be repaired, or a new reactor can be installed.

The accident has been a severe financial blow to General Public Utilities, owner of the plant. Its $300 million in insurance, the maximum amount it had been able to purchase, is far short of the $1 billion minimum that cleanup may cost. At the time of the accident, NRC ordered TMI-1 to stop operations. A number of changes were made in TMI-1, and the owners hope to be able to start operations again in 1982. With both plants closed, General Public Utilities had to buy about $14 million worth of power every month from other companies to supply its customers.

The incident at TMI-2 was investigated by several groups, including a commission appointed by President Jimmy Carter and a Special Inquiry Group appointed by NRC. Both had harsh criticisms for almost everyone involved. The major conclusion of the President's Commission was that "to prevent accidents as serious as Three Mile Island, fundamental changes will be necessary in the organization, procedures, and practices—and above all—attitudes of the Nuclear Regulatory Commission, and, to

CLOSE CALL

Monroe, Michigan

1966: About 1 percent of the fuel in an experimental reactor melted in Monroe, Michigan when the cooling system was blocked and the core partially uncovered. The reactor, a "breeder" that produced more fuel than it consumed, used liquid sodium as a coolant. A group of 25 equipment manufacturers and utility companies owned the reactor, which was named for Enrico Fermi, the physicist who led the team that built the first nuclear reactor in Chicago in 1942.

The accident started while power was being increased in the reactor after it had been shut down for some time to repair the steam generators. The reactor operators noticed that some temperatures were higher than normal and that the control rods were not as far into the core as they were supposed to be. At about the same time the reactor building's ventilation system automatically closed and isolated the building because small amounts of radiation were detected in the building. The reactor was then shut down. Radiation in the plant was so intense that it was two months before the cleanup could begin. The public, however, was not exposed to excessive radiation.

The fuel melted because a metal plate in the reactor broke loose and was carried upward by the flow of sodium coolant. It stuck on the underside of the plate supporting the fuel, blocking flow to some of the fuel. A number of the metal plates had been added to the reactor vessel as a safety measure.

The damaged core was returned to the Federal Government for reprocessing of the fuel. In 1970, the Enrico Fermi plant went back in full operation, but in 1972, it closed permanently for financial reasons. In 1982, the sodium coolant was still being stored in drums at the plant.

the extent that the institutions we investigated are typical, of the nuclear industry."

The fundamental problems, the President's Commission found, are related to people, rather than equipment. They involve the entire system that manufactures, operates, and regulates nuclear power plants. After many years of operation of nuclear plants with no evidence of harm to workers or the public, the belief that nuclear power plants are safe enough grew into a conviction. This attitude, said the commission, must be changed to one that views nuclear power as being, by its very nature, potentially dangerous; safeguards already in place must be continually questioned to determine if they can prevent major accidents.

The commission found fault with the regulations covering nuclear power plants. The regulations required that plants be designed to withstand the worst kind of equipment failure that could occur—but took no account of "less important" accidents from combinations of minor equipment failures such as those at TMI-2. Minor failures are likely to occur much more often than the worst kind of failure.

Another problem was that plant operators were not adequately trained. At TMI-2, they knew how to run the plant under normal conditions, but were unprepared for the confusing events that unfolded there. Also, operating procedures were not clear and could be read in such a way as to lead the operators to take the incorrect actions they did.

Lessons learned from previous accidents at other plants had not resulted in new instructions being passed on to plant operators everywhere, according to the commission. An engineer of the company that manufactured the steam system wrote a memorandum describing an earlier incident very similar to the one at TMI, where operators had mistakenly turned off the emergency core cooling system. Fortunately, the error did not lead to a serious accident. The engineer had pointed out that under some circumstances a very serious accident could occur. He strongly urged that other plant operators get clear instructions. But nothing was done at this earlier time.

Another shortcoming at TMI-2 was the control room. It is huge, with hundreds of alarms, and some key indicators were in places the operators could not see easily. The control room might have been adequate during normal operations, but was seriously deficient during an accident.

The commission also noted that the power company, the suppliers of equipment, and NRC were responsible for the many shortcomings they found. Therefore— whether or not operator error "explains" this particular case—an accident like that at Three Mile Island was eventually inevitable.

For about a year and a half, NRC licensed no new nuclear power plants while it concentrated its efforts on studying the accident and the lessons to be learned from it. Late in 1980, it resumed licensing of new plants issuing four new licenses in 1980 and three in 1981. The license issued in 1981 to Diablo Canyon Unit in California was suspended two months later because of questions related to its ability to withstand earthquakes.

NRC issued no new construction permits in the rest

The control room of the Unit 2 of the Three Mile Island reactor was huge and poorly designed. Its shortcomings made dealing with the 1979 accident there more difficult. ATOMIC INDUSTRIAL FORUM

of 1979, in 1980 or 1981 as it considered a long list of new safety requirements for applicants. One requirement under consideration is that new plants would have to be located farther from heavily populated areas than in the past.

Major changes were made at Three Mile Island. General Public Utilities created a new company, GPU Nuclear, to operate TMI and its other nuclear power plants. In the past, Met Ed and two other member companies of General Public Utilities had operated nuclear power plants and plants burning coal or oil. This change was recommended by the President's Commission and is intended to improve the management of the nuclear power plants.

Many changes have been made in how TMI operators are trained. They now receive more training on complex and slow-moving accidents similar to the one at TMI-2. The new courses emphasize training operators to see the "big picture," rather than just looking at a certain pump or temperature reading. The training staff has been increased from about 5 to 55 to 60, and the instructors are more highly qualified.

Changes have also been made in the control room at TMI and elsewhere. Before the accident, the control room was planned to put the maximum number of instruments in the smallest amount of space. Now, it is required that all plants design control rooms on the basis of recommendations from industrial psychologists, grouping various systems and differentiating them by color and shape.

Finally, GPU Nuclear has improved its communications with the public, which had been criticized in the investigations. In fact, Three Mile Island has become a local tourist attraction. Before the accident, about 15,000 people toured the plant's visitor center a year, while in 1981 the number was 120,000. GPU Nuclear especially

CLOSE CALL

Browns Ferry, Alabama

1975: A major fire lasting 7 hours occurred at the Tennessee Valley Authority's Browns Ferry Nuclear Plant in Alabama. At the time, two nuclear units were operating, and a third was under construction. The fire was started accidentally by a construction worker who was trying to seal air leaks with plastic foam in the electrical cable system beneath the common control room for the two operating units. He held a lighted candle near the leaks, to see if the flame flickered. Some of the foam caught on fire, and flames raced through the 1,600 cables, knocking out many of the main control systems and backup systems on both units. The plant's crew showed great ingenuity in using the remaining equipment to keep the core covered and eventually to shut the reactor down.

No workers were exposed to abnormal levels of radiation, and no radiation escaped from the plant. Some fire fighters suffered minor injuries. Cleanup required 18 months and cost $13 million. After the fire at Browns Ferry, the Nuclear Regulatory Commission recommended that nuclear power plants improve fire protection measures, improve fire control and containment, and separate and isolate safety systems from their backup systems.

welcomes nearby residents to tour the plant.

The nuclear industry also responded to the accident at Three Mile Island to improve the flow of information about abnormal behavior of nuclear reactors among power companies, reactor manufacturers, architects, engineers, universities, and Federal laboratories. The industry established the Nuclear Safety and Analysis Center to link the industry to technical experts and provide information on safety questions.

A second organization, the Institute of Nuclear Power Operations, was set up to develop "benchmarks of excellence" in nuclear power operations and to help power companies in meeting the benchmarks. In addition, the industry started an insurance company to help protect power companies against financial losses that result when a reactor cannot operate for a long period. Individual power companies reviewed their safety programs, trained reactor operators to apply the lessons of Three Mile Island, and revised some of their operating procedures.

The accident at Three Mile Island forced a reexamination of the role of nuclear energy in America's future. Opponents of nuclear power say, "We told you so." They believe that the accident confirmed the risks of nuclear power and should encourage the search for safer ways of producing energy. Supporters of nuclear energy point out that a disaster was averted, that we live with other risks created by modern technology (cars and airplanes, for example), and that we must learn from the accident and work to make nuclear power safer in the future.

IV.

Low-Level Radiation: The Invisible Problem

SINCE THE BEGINNING OF TIME, humanity has been exposed to the ionizing radiation that is part of the natural environment. Some of this radiation can damage the human body from outside while other forms damage the body when they enter it in food and drink. Natural, or "background," radiation comes primarily from these sources:

- Cosmic rays, which are streams of atomic nuclei (mostly of hydrogen) coming from the sun and stars. Because the earth's atmosphere partially screens us from this radiation, levels are higher at higher elevations than at sea level, where the protective screen is thicker. For example, a person living in Denver, Colorado, at an altitude of 1,500 meters, receives almost twice as much radiation as a person at sea level.
- Naturally occurring radioactive elements, which are

found nearly everywhere, but generally at very low concentrations. In some parts of the world—Brazil and India, for instance, where soils are rich in thorium and uranium—concentrations are higher. (Brazil nuts have 1,400 times the alpha activity of common fruits such as apples and pears, the foods with the lowest concentration of natural radioactivity.) Some naturally occurring radioactive elements, such as potassium 40, may unintentionally enter the human body in air, water, or food, and remain there. Radon 222, a gas that is produced by the decay of radium in soil everywhere, may collect in confined spaces, such as buildings. Radon is estimated to cause up to 10 percent of all lung cancer in the United States.

Some typical annual doses Americans receive from natural radiation are:

	millirems per year
Cosmic rays, sea level	29
Cosmic rays, 1,500 meters	50
Cosmic rays, air travelers (based on passengers flying 80 hours a year)	65
Cosmic rays, plane crews	160
Radioactive constituents of body	25
Gamma rays from soil (based on 25 percent of time spent outdoors)	15

From natural or background radiation, the "average" person living in the United States is exposed to about 100

ESTIMATED ANNUAL WHOLE BODY
RADIATION DOSE FOR U.S. CITIZENS

Natural Sources: 100 MREM

Medical Sources: 70 MREM

Fallout: 3 MREM
Misc.: 2 MREM
Occupational: 0.8 MREM
Nuclear Power: 0.01 MREM

millirems, distributed relatively evenly throughout the body. (The exposure of the lungs to radon 222 is not included in the "whole-body" exposure.)

Radiation from Medical Uses

Various medical activities add to the background radiation to which people are exposed. The largest single source is x rays used to search out diseased organs, decayed teeth, broken bones, or swallowed objects, as well as for other medical applications. The amount of radiation a patient absorbs during an x-ray examination depends on the purpose of the x ray and the part of the body involved. In general, the thicker the area of the body being examined, the greater the dosage required. The chest receives about 30 millirems when it is x-rayed.

X rays, unfortunately, do not uniformly travel a straight line in passing through matter, so that some radiation always escapes the target in the body at which it is aimed. To protect the body from this scatter radiation, engineers have developed cones to direct the beams to the target areas. In addition, medical personnel and patients wear lead aprons, or other devices for protection.

In 1981, about 210 million medical x-ray examinations were made, plus 100 million dental x-ray examinations. An examination may require more than one x-ray film. Some critics say that more x-ray examinations are made than are really needed. Unnecessary x rays not only pose a public health hazard but also waste money.

Starting in the 1970s, public health officials have discouraged routine use of x rays for all patients entering a hospital, or as a screening test to detect tuberculosis, breast cancer, or tooth decay. At one time, mobile vans offered a free chest x ray on the spot to anyone who walked by. This practice has almost disappeared because it needlessly exposed a large number of people to x rays for the relatively few cases of tuberculosis discovered. Even more pointless were the x-ray machines that shoe stores kept handy so that customers could determine if their new shoes fit properly by revealing the bones in their feet.

X rays or radioactive isotopes are also used to treat cancer. Because cancer cells usually divide more rapidly than normal cells, they are more sensitive to radiation. Very large amounts of radiation are directed at the cancer cells in an attempt to kill them without damaging normal cells.

Another medical use of radiation that is growing rapidly is the injection of radioisotopes into the body for diagnostic purposes. These "tracers" give off radiation that

In 1981, about 210 million x-ray examinations were made,
plus about 100 million dental x rays. The amount of radiation
a patient absorbs during an x-ray examination depends on the
purpose and the part of the body involved. A chest x-ray, for
example, delivers about 30 millirems to that part of the body.
HOSPITAL CORPORATION OF AMERICA

can be detected from outside the body by instruments
that identify where they are and the amount present. For
example, all types of iodine collect in a normal thyroid
gland, but radioactive iodine does not collect as completely in a cancerous gland. Therefore an area where
radioactivity is decreased suggests cancer may be present.
From 80 million to 100 million examinations using radioactive tracers are made every year in the United States.

These various medical uses add up to an annual radiation dose of 70 millirems for the average person living in
the United States. Obviously, some healthy people will be
exposed to no radiation from this source, while others

75

undergoing extensive medical treatment will have thousands of millirems. The 70-millirem figure merely indicates that more than 90 percent of all radiation received by the United States population from human activities is for medical purposes. The remaining 10 percent comes from radioactive fallout deposited on the earth's surface from past testing of nuclear weapons in the atmosphere, industrial use of radioactive materials, consumer products such as color TV sets and some kinds of smoke detectors, and a variety of other small sources.

Radiation from Nuclear Power Plants

About 99.99 percent of the radiation in a nuclear power plant is locked within the fuel rods. The remaining tiny fraction is collected by the plant's waste cleanup systems. A minute amount is periodically released to the environment according to standards set by the United States Environmental Protection Agency (EPA) and enforced by the Nuclear Regulatory Commission. EPA and NRC also set and enforce standards regulating the amount of radiation that workers in power plants can be exposed to.

Radioactive gases are the main source of radiation reaching the environment from a nuclear power plant. The gases are held in the plant until most of their radioactivity is lost as they decay to solid elements that can be trapped on filters. The small amount of radioactivity remaining is vented out the plant's stack. Some radiation also reaches the environment through waste waters routinely discharged from a power plant into nearby bodies of water. In addition, pipes sometimes leak, filters fail, and workers open the wrong valve. All activities of nuclear power plants add 0.01 millirems to the annual dose of radiation the average American receives. People living within 80 kilometers of Three Mile Island received an average of under 2 millirems from the accident there, according to estimates prepared by scientists from several Federal agencies.

These radiation exposures are for the general public. Some people have higher exposures, including patients receiving a large number of x rays or being treated with radiation, uranium miners, people working with radioactive materials in industry and universities, and employees in nuclear power plants.

Biological Effects of Radiation

The effects of ionizing radiation are better known than the effects of many other toxic materials, although our knowledge is still far from complete. Some of this knowledge comes from tests on laboratory animals and some from human experience with radiation. Among the people who have suffered large doses of radiation are the Japanese survivors of the atomic bombs dropped on Hiroshima and Nagasaki, the few people injured in radiation accidents, and people who studied and worked with radioactive substances for long periods of time before the hazards of radiation were understood. For example, Marie Curie, the chemist who discovered the radioactive elements radium and polonium, died in 1934 of a form of anemia probably caused by the materials she worked with.

Pioneers in the use of radiation in medicine in the

1920s were exposed to far higher doses than is the case with members of their profession today. Some industrial workers also were exposed to large amounts of radiation.

The best known example is the workers in the watch industry in the United States and Switzerland who brushed radium paint on the dials of watches, clocks, and instruments to make them glow in the dark. To get a fine point on their brushes, they touched the tip to their lips or tongues, and so swallowed small amounts of radium that later proved deadly to many of them. The radiation doses

These factory workers painted radium paint on the dials of watches in the 1920s. To get a fine point on their brushes, they often touched the tips to their lips or tongues. The radium that entered their bodies as a result later proved deadly to many of them. ARGONNE NATIONAL LABORATORY

that caused death in these cases were generally high, and it is impossible to predict with certainty from experience with high doses how low doses affect the body.

Ionizing radiation clearly can cause cancer and do damage that may be passed on to offspring. What is uncertain is the relationship between the size of the dose— particularly doses below 1 rem—and the harm that results.

Several factors make it difficult to relate cause and effect. It can take 25 years or more after exposure for cancer to develop and a generation or more for hereditary damage to become evident. In both cases, other causes are possible. Cancer, for instance, has been linked to cigarette smoking, certain industrial chemicals, foods, alcohol, viruses, ultraviolet light, and perhaps heredity. Also, some individuals—particularly children and an unborn child growing in the mother's uterus—are extra sensitive to radiation because of the large number of rapidly growing cells in their bodies. In addition, certain parts of the body are particularly susceptible to damage from radiation. For example, bone marrow cells, which are constantly producing blood cells, are more sensitive than muscle and nerve cells, which grow more slowly.

Public Health Standards

The Environmental Protection Agency is responsible in the United States for setting the broad public health and environment standards for the nuclear cycle as a whole, while the Nuclear Regulatory Commission handles the detailed regulation of individual plants under those standards. The exposure to radiation that the standards permit has been lowered continually since the Federal Government first established limits in 1957. In 1977, EPA lowered the radiation dose to the public from the nuclear fuel cycle from 170 millirems to 25 millirems a year. EPA did not claim that this level is free of risk. Instead, the level balances the risk of health effects against the costs of controlling releases of radioactive materials.

Radiation standards are based on the principle that the number of mutations to cells is directly proportional to the radiation dose absorbed—that is, if the radiation dose increases 10 times, the number of mutations increases 10 times. Thus EPA accepts the assumption that there is no "threshold" for the mutational effects of radiation; no dose is so low that it is absolutely harmless.

The principle of a threshold differs for somatic (body) cells and reproductive cells. A somatic cell may suffer a mutation from low levels of radiation, leading to cancer. However, if a small radiation dose affects growth and prevents some somatic cells from dividing, the remaining unaffected cells can take up the slack. If few cells are affected, there may be no permanent damage.

Reproductive cells, on the other hand, function alone. Once it has been fertilized, a damaged egg cell cannot be replaced by an unaffected cell. Suppose only one reproductive cell out of 1 million is damaged. Then it will take part, on the average, in one out of 1 million fertilizations. And once it is involved, it does not matter that 999,999 undamaged cells are still available—it is the damaged cell that is fertilized and reproducing. That is why there is no threshold to the hereditary effects of radiation and why there is no "safe" amount of radiation insofar as hereditary effects are concerned. However small the quantity of radiation absorbed, there is a corresponding increase in the overall burden of undesirable reproductive cells (as well as of undesirable somatic cells).

Shortly after the accident at Three Mile Island, EPA received a report from a committee of the National Academy of Sciences, an organization of prominent scientists that frequently advises the Federal Government on science and technology. EPA had asked the academy, which had evaluated the risks of radiation exposures in 1972, to reevaluate them. The academy's report illustrates how deeply divided scientists are regarding the risks of low levels of radiation. The report concludes that there are some exceptions to the "no threshold" rule, but none significant to safety standards. However, 6 of 21 members of the committee contend that there is a threshold below which injuries are much less frequent and possibly don't exist at all. They claim that some of the information in the report is incorrect and that it contributes to unnecessary fears about radiation hazards.

The question of low levels of radiation is not new. From the end of World War II until the early 1960s, debate over it was intense. The debate centered on the testing of atomic bombs in the atmosphere, but there were also warnings about the dangers from peaceful uses of nuclear energy. The debate died down after 1963, when the United States and the Soviet Union agreed to stop atmospheric testing. In addition, government officials and scientists seem confident that radiation from the nuclear power industry could be kept at a safe level.

By the 1970s, the debate heated up again for a number

of reasons. One was the disclosure that in the 1950s and 1960s the United States government had not publicized reports showing harmful effects from low levels of radiation. In addition, several studies suggested that nuclear workers exposed to radiation were developing cancer at higher-than-normal rates.

The general distrust in government agencies that followed the Vietnam War and Watergate scandals contributed, as did the fears of many scientists that accidents would become more likely as more nuclear power plants were built. Three Mile Island brought the question of low-level radiation into the headlines once again and into the national debate on the future of nuclear energy.

V.

Radioactive Wastes: The Future Problem

AT PRESENT, the only practical means of reducing the radiation coming from nuclear wastes is to allow them to decay naturally. Each radioactive element decays and releases heat and ionizing radiation at its own particular rate and continues to do so no matter what is done to it.

In the 1960s, nuclear experts were optimistic that the problems of disposing of radioactive wastes could be managed. As a result, little money was put into the search for long-term solutions, and what now appear to be shortcuts were occasionally taken in how wastes were treated, stored, or disposed of. This past neglect, along with the general distrust of government and large-scale technology, makes solving the radioactive waste problem far more complicated than it was in the 1960s. The entire future of nuclear power has become the subject of intense debate. The nation continues to struggle to develop an overall policy for permanent disposal of radioactive wastes, while at the same time, the quantities requiring disposal are rising rapidly as more nuclear plants go into operation.

Sources of Wastes

Radioactive wastes are produced by all the various steps in the production of nuclear electricity, nuclear weapons, and nuclear fuel for ships, as well as the use of radioactive materials in medicine, research laboratories, and industrial operations. Well over 90 percent of all nuclear wastes now in existence come from national defense activities. Not all nuclear wastes are the same:

Low-level wastes are generated almost any place where radioactive materials are handled. They emit primarily small amounts of gamma and beta radiation. Low-level wastes include rags, protective clothing, hand tools, and similar items used around nuclear reactors; vials, needles, test tubes, and other materials used in medical, research, and industrial activities; and solutions used in rinsing and other cleanup operations.

High-level wastes give off large amounts of ionizing radiation and heat. Presently, almost all of the wastes in the high-level category were produced as liquids during manufacturing of nuclear weapons and reprocessing of spent fuel from United States Navy nuclear reactors. The wastes contain fission products, plus some transuranic elements (synthetic elements such as plutonium that are heavier than uranium and are radioactive). Initially, beta and gamma radiation are responsible for most of the radiation from these wastes. Because of the intense gamma rays, heavy shielding is required. After 500 to 1,000 years, alpha rays present the greatest hazard.

Transuranic wastes are solids contaminated with elements heavier than uranium to such an extent that they should not be disposed of at the surface of earth. Some transuranic elements—plutonium 239 for example—have very long half-lives and are toxic. Transuranic wastes exist in many of the same forms (rags and tools, for example) as low-level wastes. They can be generated by almost any process involving transuranic elements and by reprocessing of spent fuel, but all now in existence come from production of plutonium for nuclear weapons. The primary radiation is alpha particles. Thus no shielding is required, although it is desirable to contain them, transuranic wastes are less radioactive and generate less heat than high-level wastes. Still, because of their long half-lives and toxicity, transuranic wastes require careful permanent disposal. They are generated in large quantities, so that the methods used for the smaller quantities of high-level wastes are not always practical.

In addition, two other types of materials can be considered radioactive wastes: *spent fuel*, most of which is stored at nuclear power plants, and *tailings*, the solid materials from the mining and milling of uranium. Tailings contain low levels of natural radioactive elements that emit alpha, beta, and gamma radiation.

Tailings

The processing of uranium ore to make yellowcake leaves a mixture of sand, chemicals added during processing, and waste water. The mixture is generally piped to holding ponds, where it is contained by a series of dikes sometimes made of the tailings themselves. In the past, a pond was allowed to dry. The tailings were removed and piled up elsewhere at the mill, and the pond was ready for use again. Currently, the tailings are left in a pond and the area is reclaimed after the pond is dry.

Tailings consist of finely divided particles that are easily moved by wind and water. The particles contain about 0.1 percent of uranium, most of it uranium 238, which decays over 4.5 billion years to become lead 206, a stable nonradioactive element. The long process passes through a series of intermediate elements called decay products; these, too, are radioactive. The greatest hazard comes from radon 222. Unless covered with thick layers of soil or other materials, tailings may release the gas into the atmosphere. Radon 222 has a short half-life (under

Piles of uranium tailings are wetted down to prevent the fine particles from being blown away. ATOMIC INDUSTRIAL FORUM

four days). However, because it is formed by the decay of radium 226 (which has a half-life of 1,600 years), new supplies continue to be created. Some nuclear experts fear that because radioactive materials from uranium tailings can so easily find their way into air, water, and the food chain, tailings may be more dangerous than high-level wastes, all of which are well isolated from the environment.

Uranium tailings were largely ignored in the early days of the nuclear era. The Atomic Energy Commission (AEC), the Federal agency that was responsible for developing, promoting, and regulating United States nuclear activities from 1946 to 1974, claimed it had no jurisdiction over that part of the nuclear fuel cycle, even though all uranium up to 1970 was produced for the Federal Government. As a result, when a uranium mill closed, the tailings were abandoned and left unprotected. By the early 1980s, 140 million metric tons had accumulated at active and inactive mills in 11 states, most of them in the West.

Not all tailings piles are in out-of-the-way places. One, for example, is located 30 blocks from Utah's State capitol building in Salt Lake City. In 1964, Vitro Chemical Company, which is now out of business, left 1.7 million metric tons of tailings there when it closed its mill.

Another pile is near Grand Junction, Colorado. Starting in the 1950s, about 3,000 homes and other buildings in and around the city were built using concrete containing uranium tailings, exposing people in the buildings to high doses of radon gas and other radioactive materials. (Radon can also accumulate in buildings made of naturally occurring rocks if the buildings are well insulated and weatherstripped, which minimizes movement of air in and out.)

The Federal Government and the state of Colorado have agreed to spend $22 million to remove the radioactive materials from the 800 structures with the highest levels of radiation. The job will require several years to complete. Tailings also have been used in concrete mixtures for buildings in other Western cities, but not on as large a scale as in Grand Junction.

In 1978, Congress passed the Uranium Mill Tailings Radiation Control Act. Under the act, the Federal Government is to clean up 24 million metric tons of tailings at 24 abandoned mills where uranium had been processed under contracts with the Atomic Energy Commission, and at one site, Canonsburg, Pennsylvania, where other metals had been processed. First on the list are piles where public exposure is greatest—Salt Lake City, Utah; Durango, Colorado; Canonsburg, Pennsylvania; and Shiprock, New Mexico. If release of radiation from the piles can be significantly reduced by covering them with soils, the Government will buy the land and cover the piles. If they can't be controlled where they are, the tailings will be removed to more suitable Federal land, where they will be covered. Many uranium mills are located on river banks because they use a lot of water. Any abandoned tailings piles subject to flooding may have to be removed.

The act also requires that piles at working uranium mills be controlled. In the early 1980s, there were tailings piles at about 25 active uranium mills in the United

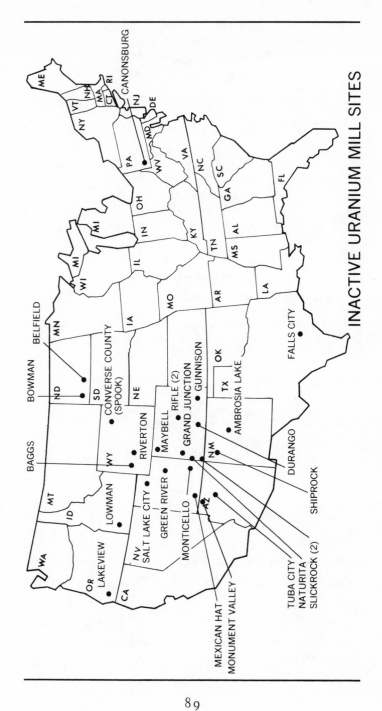

INACTIVE URANIUM MILL SITES

Inactive Uranium Mill Sites
Scheduled for Remedial Action

PRIORITY	High Priority Sites
HIGH	Salt Lake City (Vitro), Utah
	Canonsburg, Pennsylvania
	Durango, Colorado
	Shiprock, New Mexico
	Grand Junction, Colorado
	Riverton, Wyoming
	Gunnison, Colorado
	Rifle, Colorado (2)
	Medium Priority Sites
MEDIUM	Mexican Hat, Utah
	Lakeview, Oregon
	Falls City, Texas
	Tuba City, Arizona
	Naturita, Colorado
	Ambrosia Lake, New Mexico
	Low Priority Sites
LOW	Green River, Utah
	Slick Rock, Colorado (2)
	Maybell, Colorado
	Monument Valley, Arizona
	Lowman, Idaho
	Converse County (Spook Site), Wyoming
	Baggs, Wyoming
	Belfield, North Dakota
	Bowman, North Dakota

States. The largest, near Grants, New Mexico, contains 23 million metric tons of tailings. It covers more than 1 square kilometer and rises as high as 30 meters. In controlling tailings at uranium mills in the future, emphasis will be on locating new mills in remote areas and disposing of the tailings where there will be no erosion. The most practical method probably would be to place the tailings into the open pits created when the ore was mined. This could require additional measures to make certain that radioactive materials do not seep out and contaminate surface or ground water, which is the source of springs and wells.

The tailings are not the only concern in uranium mining and milling. The ponds where the tailings are first stored also must be managed properly to prevent the escape of radioactive materials. In 1976, radioactive wastes spilled out of a pond at a uranium mill in Church Rock, New Mexico, after a crack developed in the pond's wall. In the hour it took workers to seal the crack, about 350 million liters of tailings solution and 1,000 metric tons of tailings sand gushed into a small stream. Small amounts of radioactivity were found as far as 120 kilometers into Arizona, making Church Rock the first spill of radioactive wastes in the United States to escape the mill site.

Low-Level Wastes

In the 1940s and 1950s, low-level wastes were buried in shallow pits at sites operated by the Atomic Energy Commission. Small amounts also were placed in drums and dumped at sea by the United States and other countries as well. The United States stopped dumping at sea in 1970. The Federal Government now operates 14 burial grounds for wastes produced by its nuclear facilities.

Until the early 1960s, AEC permitted the relatively small amounts of low-level wastes then being produced by

Wastes with low levels of radioactivity (such as the chemical processing equipment shown here) can be safely disposed of in shallow pits at special burial sites. BATTELLE NORTHWEST LABORATORY

commercial sources to be buried at its sites in Oak Ridge, Tennessee, and Idaho Falls, Idaho. Later, six sites for burial of commercial low-level wastes were opened by private companies, the first in 1962, in Beatty, Nevada. Sites were opened later in Barnwell, South Carolina; Richland, Washington; Maxey Flats, Kentucky; Sheffield, Illinois; and West Valley, New York.

In the late 1970s, problems emerged at a number of these sites. The West Valley site was closed because poor drainage caused the burial trenches to fill with rainwater and overflow. When Kentucky state officials discovered that some radioactive materials had seeped out of the Maxey Flats site, they placed such a high tax on wastes buried there that the operation soon became uneconomical and closed. The Sheffield site closed after it became filled and the Illinois governor would not permit it to expand.

In the West, the governors of Nevada and Washington closed sites in their states temporarily because trucks were delivering damaged and leaking containers. While the western sites were closed, some hospitals and research laboratories were forced to reduce their work with radioactive materials because there was no place to dispose of the low-level wastes.

If the burial sites in Washington, Nevada, and South Carolina continue to operate, they should be able to handle all low-level wastes generated in the 1980s. These states, however, resent being a dumping ground for wastes from other regions. A Federal law passed in 1980 encourages states to cooperate in setting up a burial site for all of them to use. The first agreement was signed in 1981 by Washington, Oregon, Idaho and Utah. They will send their wastes to the existing site in Richland. Regions without an existing site are likely to find it more difficult to reach an agreement.

Transuranic Wastes

Before 1970, transuranic wastes were not always separated from other wastes contaminated with low levels of radioactivity. Then the Government decided that transuranic wastes, because of their long half-lives and toxicity, should no longer be buried in shallow pits with low-level wastes. Now such wastes are stored in drums and boxes that can be retrieved later for further process-

ing, if necessary, and permanent disposal. Most transuranic wastes are stored by the Federal Government at
the Hanford Reservation near Richland, Washington,
the National Engineering Laboratory in Idaho Falls, and
the Savannah River Plant in Aiken, South Carolina. The
Federal Government is designing an incinerator that
would reduce the wastes to a rocklike slag and also developing a process to bind the wastes into glass, ceramics,
or other materials that would be easier to dispose of.

Low-Level Wastes Going to Commercial Burial Grounds Come from Several Sources (1978)

Nuclear power plants	43%
Medical and research facilities	25%
Industry	24%
Government and military operations	8%

High-Level Wastes

Almost all present high-level wastes result from defense activities. The first wastes generated (in 1944) were
stored in large steel-lined concrete tanks buried at the
Hanford Reservation, which was producing plutonium for
nuclear weapons. The tanks, built hurriedly during
World War II, were made of a single wall of carbon steel,
which corrodes easily if it is in contact with acids. Stainless steel, which is less likely to corrode, was not available
because of the war. The wastes, which were acidic, were

neutralized before being pumped into the tanks. This changed the radioactive materials in the wastes to solid particles and also helped reduce corrosion. Additional tanks with double walls were built later at the Government's Savannah River plant in South Carolina and the National Engineering Laboratory in Idaho Falls. The tanks in Idaho are of stainless steel, which permits acidic wastes to be stored without treatment. In the early 1980s, about 288 million liters were stored at three locations:

> 186 million liters at Hanford
> 91 million liters at Savannah River
> 11 million liters at Idaho Falls

At first, the Government predicted that the wastes at Hanford could be safely stored for decades, possibly centuries. But by the mid-1950s, some of the Hanford tanks were leaking. One leak in 1973 went undiscovered for 55 days. In all, about 435,000 liters of radioactive wastes leaked into the ground at Hanford. The soil at Hanford is dry and porous and so was able to soak up the wastes. Thus far, there has been no contamination of ground-water. Radioactive wastes were even dumped directly into the ground at Hanford. The soil there is so dry and porous and groundwater is located so deep, that Hanford officials approved the dumping.

Once the tanks began to leak, the Atomic Energy Commission took various steps to improve the methods of storing high-level wastes. New tanks were built with double walls and improved devices to detect leaks. Wastes in the stainless steel tanks at Idaho Falls were converted to dry granular particles, which were stored in

The storage tanks at the Federal Government's Savannah River Plant in South Carolina are built of carbon steel surrounded by concrete about 1 meter thick, set about 13 meters in the ground and covered with dirt. Pictured here are two steel tanks before concrete encasement. E. I. DUPONT DE NEMOURS & CO., S.C.

underground steel bins inside concrete vaults. The solids can be retrieved from the vaults if necessary. Most of the water in the neutralized wastes at Hanford and Savannah River was evaporated, leaving a damp cake in the tank.

According to a report from the National Academy of Sciences, the Federal Government faces a problem "on an enormous scale" in permanently disposing of its high-level wastes. Critics of nuclear energy consider the government's record of managing high-level wastes dismal. Federal officials point out that very little of the wastes leaked, that the small amount that did was from old, improperly built tanks, that no one has been exposed to radiation, and that there is no risk of such exposure in the future.

Several methods are being studied to convert the wastes

to solids that will withstand the intense radiation and high temperatures generated by the high-level wastes themselves, as well as possible contact with water. In "surface entombment," the wastes would be removed from the storage tanks, treated to produce ceramic-like particles, mixed with cement, and pumped back into the tanks to solidify. The tanks would then be covered with a meter or more of concrete. The wastes might also be solidified in copper containers with walls about 5 centimeters thick; the containers would then be buried deep in rock formations.

Spent Fuel

Much of the disagreement about radioactive wastes involves wastes from commercial nuclear power plants. Such plants are large and highly visible (in contrast to plants making nuclear weapons), and citizens have a right to participate in public hearings on their regulation.

For years, the general plan was for commercial reprocessing plants to convert the unfissioned uranium and plutonium in spent fuel into new fuel. Reprocessing plants, which must be large to be economical, were expected to be built once enough power plants were in operation to keep them busy.

In 1976, President Gerald Ford temporarily postponed commercial reprocessing and a year later President Jimmy Carter "indefinitely deferred" reprocessing. Both presidents feared that reprocessing would greatly increase supplies of plutonium in a form that would lead to the spread, or "proliferation," of nuclear weapons. At about that time, India had exploded a nuclear bomb made out of plutonium from reprocessing the spent fuel from a small reactor it had purchased from Canada. Also some industrial nations were considering selling reprocessing plants to other less industrialized nations.

Reprocessing also would increase the opportunities for terrorists to steal plutonium, which is now held under the strict control of the Federal Government. Opinions differ as to whether a few terrorists would have the skills to build an atomic bomb. Most nations probably could do so, however, if they were willing to devote large sums of money and effort to it.

In 1981, President Ronald Reagan reversed the Carter policy and, once again, private industry will be permitted to reprocess spent reactor fuel. At the time, however, no commercial reprocessing plants were in operation in the United States, and there were no prospects for any in the forseeable future.

United States industry has already made several unsuccessful attempts to get into the reprocessing business. In 1966, Nuclear Fuel Services began reprocessing spent reactor fuel in West Valley, New York, at a plant that also handled low-level wastes. The plant was the first commercial plant in the United States to reprocess fuel and handle high-level wastes. Between 1966 and 1972, the plant reprocessed 619 metric tons of spent fuel, 375 metric tons from Hanford and 244 from private power companies. Then the plant was shut down to increase its capacity and to make changes that would reduce the high levels of radiation to which workers were being exposed. It never reopened. It left behind two buried tanks. One

This plant, operated by Nuclear Fuel Services, Inc. at West Valley, New York, was the first privately owned plant in the United States to process wastes with high levels of radioactivity. When the plant was shut down in 1972, large amounts of liquid radioactive wastes were left there, as well as spent fuel rods. Solidifying and disposing of the wastes may cost as much as $300 million. W. R. GRACE AND COMPANY

was filled with 2.3 million liters of high-level wastes. A second tank contained 450,000 liters of liquid waste from reprocessing thorium, which was being studied for conversion to reactor fuel. Also left behind was a pool holding 750 spent fuel rods from commercial power plants.

At the end of 1980, the company's lease on the site expired, creating legal problems as to who would have to clean up the plant. The company? Or New York State, which had leased the site to the company, gave the company money to build tanks for storage of high-level wastes, and set up a perpetual care fund (to which the company also contributed)? Or the Federal Government, which had encouraged construction of the plant and was its biggest customer? In 1980, Congress authorized the Federal Government to develop a plan for solidifying and disposing of the wastes. Under the plan, the Federal Government is to pay most of the cost, estimated at between $200 million and $300 million, and New York is to pay the rest. Cleanup is scheduled to be completed in the 1990s.

Another unsuccessful attempt at commercial reprocessing took place in Morris, Illinois, where General Electric Company built a plant in the early 1970s. Because the process the company chose didn't work properly, the plant never went into operation. Instead, it is now storing spent fuel shipped there from several power companies.

The Illinois attorney general is opposed to expanding the site when it is full, and so are some citizens of Morris. They fear that their town will become a national warehouse for nuclear wastes until a method for permanent disposal is developed, which probably won't be until the late 1990s. When spent fuel started arriving in Morris in the 1970s, there were no objections. Now, some people in Morris share the growing national concern about what to do with nuclear wastes.

Politics was behind the failure of another company to reprocess spent fuel. Allied General Nuclear Services started building a plant at Barnwell, South Carolina, in the 1970s. In 1977, when President Jimmy Carter in-

*In the 1970s, Allied General Nuclear Services started building
this plant to reprocess spent fuel from nuclear power plants.
Because of uncertainties about the future of reprocessing in
the United States, the company stopped work on the plant
in 1981 after investing more than $362 million in its
construction. DOE* PHOTO

definitely deferred commercial reprocessing of spent fuel,
plans for the plant came to a halt.

By 1981, when President Reagan lifted the ban on
commercial reprocessing, the company had invested more
than $362 million in the plant. Shortly thereafter the
company decided to close the plant unless some one
would buy it. The company fears that some future President might again ban commercial fuel reprocessing and
believes that reprocessing will have to be done by the Federal Government, as it is now in several other countries.

President Reagan, however, thinks that industry should
conduct its business with as little help as possible from
the Federal Government, and he is also reducing Federal
spending. Other companies are not eager to get into the
reprocessing business because at present it is cheaper to

make fresh fuel from uranium than to reprocess spent fuel.

Except for the small amounts at West Valley and Morris, all spent nuclear fuel is stored in swimming pools at the power plants where it was used. In 1981, 6,500 metric tons were in storage. The pools were built to last until reprocessing plants began operating. With no plants operating, the pools are rapidly filling up. In 1977, a power plant at Hartsville, South Carolina, ran out of storage space and had to arrange for storing some of its spent fuel at another of its nuclear power plants. Other plants may soon face the same problem. Almost all have applied to the Nuclear Regulatory Commission for licenses to pack the fuel assemblies closer together or to expand their pools. Granting the licenses would take care of the spent fuel for perhaps a decade. Some states and local communities, however, oppose any expansion.

The question of reprocessing will have an important effect on future disposal of high-level wastes. In the opinion of some nuclear experts, it is simpler to dispose of spent fuel assemblies than to reprocess them because they are already in the form of a ceramic. These experts believe that there is greater risk in reprocessing spent fuel and solidifying the waste liquid formed during reprocessing. In the opinion of other experts, burying spent fuel is foolish. It represents energy potential equal to billions of barrels of oil. Our nuclear fuel resources are limited and should be conserved by recycling them. Experts also disagree about how much burial space each method would require. Thus, as in so many other questions in the nuclear debate, there seems to be no agreement.

Officials of a nuclear power plant in Hartsville, South Carolina, stand at the end of the railcar holding a shipping cask filled with spent fuel. In 1977, the plant ran out of storage space and had to arrange for storing some of its spent fuel at another nuclear power plant. ATOMIC INDUSTRIAL FORUM

Permanent Disposal

The need for developing a safe and secure way for permanent disposal—or "long-term isolation"—of radioactive wastes has been recognized since the very beginning of the nuclear age. Burial has always been the most obvious method, but scientists have suggested a number of others.

One is to deposit the wastes in Antarctica. A container of waste would be placed in a shallow hole drilled into the ice there. As the ice melted from the heat given off by the wastes, the container would slowly sink. The melted ice would refreeze above the container as it sank, sealing it off. The idea was dropped because of the international disagreements that would have arisen and also because not enough is known about the stability of icecaps over the thousands of years required for wastes to decay.

The National Aeronautics and Space Administration (NASA) has proposed disposing of wastes in space. A container of waste would be lifted by a space shuttle into orbit around the earth. The container would then be transferred to an unmanned space vehicle, which would be launched into orbit around the sun. There appears to be no fundamental scientific barrier to space disposal, but many technical questions remain to be solved. The containers would have to be able to withstand an explosion on the launch pad or failure of the rocket. The risks, plus the very high cost, mean that space disposal will likely be nothing more than an interesting idea.

Another proposal is similar to attempts of the ancient alchemists who wanted to create gold and silver from less valuable metals. The alchemist of the twentieth century would bombard radioactive elements with neutrons in a reactor to change them into shorter-lived or stable elements. The problem is that present reactors are not very good at converting most fission products and transuranic elements into less harmful substances.

One proposal made some years ago is still being studied: disposing of wastes under thick red clay deposits that lie on the seabed in the center of large ocean basins. These deposits are about 30 meters thick and lie 500 meters or more below the surface of the ocean. They are very stable; some appear to have been undisturbed for 70 million years. The sites have almost no plant or animal life or valuable natural resources that might encourage exploration in the future. One means of depositing the wastes would be to load them into a bullet-shaped container. When dropped into the ocean, the container would penetrate the deposits, which would then flow in behind the container to seal up the hole.

Many questions remained to be answered. One is the possible effects of the waste's heat and radiation on clay. The Federal Government is studying the engineering problems involved in placing containers in the clay deposits and, if necessary, retrieving them. It may be 1990 before this way of disposing of radioactive wastes has been studied adequately. Japan and Great Britain, which have few sites for disposing of their radioactive wastes on land, are seriously considering disposal at sea, which is open to all nations. If disposal at sea proved feasible, it would offer a solution to the enormously complex social and political problems associated with other alternatives, which always seem to stir up conflict among individuals, communities, states, and nations.

Another proposal calls for depositing radioactive wastes in a very deep hole. The surrounding rock would contain the wastes and the great depths would delay the release of radiation into the human environment. Three factors are important to this concept. The geology deep in the earth is largely unknown. Deep holes are expensive and difficult to drill, although it is now possible to drill a narrow hole to 10,500 meters, which should be deep enough. Finally, lowering the waste container on a wire to that depth could pose severe engineering problems. Much more information is needed to evaluate this proposal.

Delayed disposal is the solution recommended by some scientists. Spent fuel or solidified high-level wastes would be stored in pools or air-cooled vaults. During storage, the radioactivity and heat would decrease to levels that would make the waste more manageable. France has chosen this route. After solidified high-level wastes have been stored in underground vaults for about 50 years, they will be buried at far greater depths in a suitable rock formation.

Opponents of nuclear power want a decision on permanent disposal before the nuclear power industry is allowed to expand further. As things are working out, delayed disposal may end up being part of the United States strategy for managing radioactive wastes, not because of technical factors but because of political realities. It appears highly unlikely that a method for permanent disposal will be in operation for at least a decade. So for most spent fuel now existing in the United States, permanent disposal almost certainly will be delayed.

Geological Disposal

With present technology, "geological disposal" appears to be the surest way of safely isolating nuclear wastes for long periods of time. The wastes would be buried several hundred meters below the surface of the earth in stable rock formations. From the surface, a nuclear waste "repository" would resemble a relatively large mine. There would be a railroad siding; facilities for deep excavation; and buildings for unloading, handling, and repackaging nuclear wastes before they are buried. Where necessary, concrete shielding would protect workers from radiation. Ventilation systems would be filtered to prevent release of radioactive particles or gases into the area. Several vertical shafts would lead from the surface to tunnellike storage areas about 800 meters deep. The wastes would be placed in holes excavated along the tunnels. As each storage zone filled, the holes, tunnels, and shafts would be filled in and sealed. However, for some time prior to

Storage of spent fuel 420 meters below the surface in granite rock is being studied at the Nevada Test Site. The spent fuel is stored below the circular covers in the center between the rails used by the vehicle that transfers the fuel in heavily shielded casks. The electrical heaters at the sides are being used to determine how granite responds to the heat that may be present in an actual repository for high-level wastes. This will determine if radiation reduces the ability of the rock to hold high-level wastes.

final closure, the wastes could be retrieved if problems developed or better disposal methods became available.

Geological disposal involves two factors: where to bury the wastes and how to "package" them.

A site for geological disposal must meet several requirements:

- It must be a stable geological environment, one that can prevent wastes from wandering from where they are placed, and one that can be expected to change little in the future.
- It should withstand heating and radiation caused by the decay of radioactive elements.
- It should contain as little water as possible because water is the primary way that nuclear wastes can escape the disposal site.

One likely type of rock formation, singled out by the National Academy of Sciences in 1957, is the salt that lies in thick beds or dome-shaped deposits under some parts of the United States. Salt has many desirable features. Because they are very soluble in water, salt deposits cannot contain many cracks through which water can move or they would dissolve. Also, because salt flows when it is heated under pressure, any cracks that do develop tend to seal themselves. Finally, the age of many deposits—200 million or more years—demonstrates that they are stable and isolated from circulating groundwater.

As scientists studied salt formations, however, they discovered some problems. Over thousands of years, some

groundwater may have worked its way into salt formations and changed them in ways that are hard to detect
from the surface and are not yet fully understood. Small
pockets of salt water have been found in some salt formations. These pockets could burst when heated by the
decaying wastes. To avoid corrosion from the salt water
produced, waste containers would have to be made of
special alloys. Also, salt deposits often lie near valuable
minerals, including natural gas, petroleum, and gypsum.
Thus there is a chance that future prospectors in search
of natural resources might accidentally drill into buried
wastes if their locations are lost over the centuries.

Because so much seemed known about salt deposits,
the Atomic Energy Commission, in 1970, selected an
abandoned salt mine near Lyons, Kansas, as a place to
build a repository to test disposal of high-level wastes.

AEC had to give up the idea in 1972, when environmental groups and Kansas State geologists discovered serious flaws in the site that had escaped the attention of
Federal planners. Less than 800 meters away from the
proposed site was an active salt mine whose owners had
previously experimented with water as a way of cracking
the salt deposits. In one experiment, the mine operators
"lost" 600,000 liters of water. No one could be sure
where it had gone, and it might turn up in the nuclear
waste area. In addition, numerous old wells drilled in
search of mineral deposits were discovered nearby. Thus
there were too many uncertainties about conditions deep
in the earth where the wastes were to be deposited.

The Government then proposed building a repository
in solid rock resembling granite 500 meters or more below

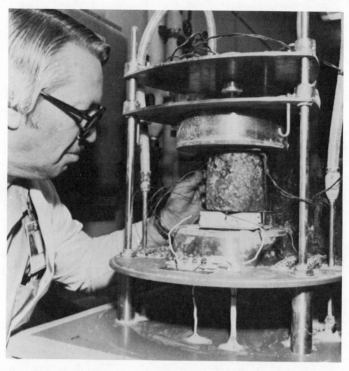

A Federal scientist places one of several thermocouples—
heat-measuring devices—in a small hole drilled into a cylinder
of bedded salt. Thermocouples are placed at various spots
along the side of the cylinder to help determine how rapidly
heat travels through salt. The bottom of the salt sample was
heated to temperatures in excess of 370° C. These tests
provide information needed to perform complex calculations
of how salt behaves. SANDIA LABORATORIES

the Savannah River Plant, where geological conditions
were favorable. Because of the opposition of a United
States Senator from South Carolina, Congress never ap-
propriated money for the project.

Federal policy then shifted to building a facility near
the surface of the earth to store wastes for an indefinite

period, and retrieve them when a safer repository was built. Three years later, that plan was set aside and salt was back in favor. The Government selected a site in northern Michigan, but backed down in the face of opposition by local citizens.

In 1975, still searching for a suitable salt formation for a repository, the Government announced plans to build a "Waste Isolation Pilot Plant" on Federally owned land near Carlsbad, New Mexico. The plan was changed several times because of political opposition, both in New Mexico and in the United States Congress.

Then in 1980, President Carter announced an overall Federal policy for nuclear wastes, which canceled plans for the Carlsbad plant and identified 11 sites for possible construction of a repository, among them Carlsbad. Nine underground salt domes were selected: eight along the Gulf Coast in Mississippi, Texas, and Louisiana, and one in Utah. The two remaining sites were already committed to Federal nuclear activities: the Hanford Reservation and the Nevada Test Site, where the Government conducts underground nuclear tests. A site is scheduled to be selected by 1987, and a repository is to be in operation in the late 1990s at the earliest.

Studies at Hanford and the Nevada Test Site are to evaluate rock formations other than salt, which might permit building storage sites in several locations and so minimize the transport of wastes across the country. Although more is known about salt formations, they do not have a clear technical advantage over a number of other rock formations.

Types of rocks under consideration include granites

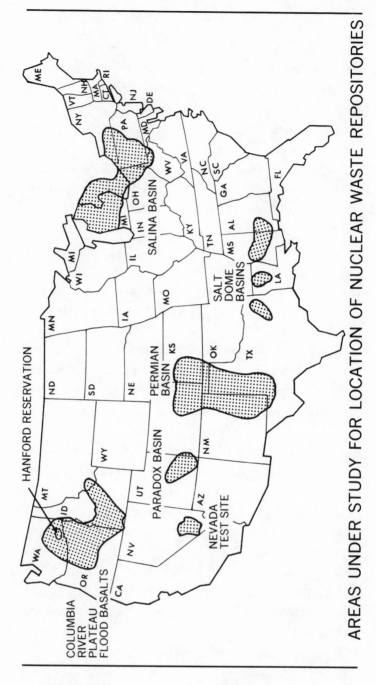

AREAS UNDER STUDY FOR LOCATION OF NUCLEAR WASTE REPOSITORIES

and basalts, which are formed when molten rock cools, and tuffs, which are solidified volcanic ash. Unlike salt, all of these rocks are likely to have some cracks. If the cracks are connected to one another, water can pass through the rock. But unlike salt, these rocks can chemically bond with most radioactive elements.

Most studies suggest that the odds are very low that ground water will provide a route for radioactive materials to escape a repository built in a suitable rock formation because the half-lives of many radioactive elements are short and others would be trapped by surrounding rocks. There are, of course, other ways a repository could be damaged and release radioactive materials—human activities such as digging, drilling, or sabotage, as well as natural events such as earthquakes. Any site selected, however, would be highly unlikely to be subject to any such natural catastrophes.

An open pit uranium mine in Oklo, a village in Gabon, Africa, provides evidence that nuclear wastes can be disposed of safely in stable rock formations. About 1.8 billion years ago, the unusual conditions necessary for a chain reaction were established in a rich vein of uranium ore. After a portion of ore was consumed, this natural nuclear reactor shut down. It left in the vein the same

Geological exploration has concentrated on four regions underlain by salt deposits—the Salina Basin, the Paradox Basin, the Permian Basin, and the Salt Dome Basins on the Gulf Coast—as possible locations for repositories for high-level radioactive wastes. More recently, basalts in Washington and Nevada have begun to be studied.

kinds of fission products that a modern reactor produces, an unmistakable sign that a chain reaction took place. Studies show that the vein has been remarkably stable since the chain reaction took place over 2 billion years ago.

Packaging Wastes

The second crucial part of the national strategy for permanently and safely disposing of nuclear wastes is the development of methods to package them—that is, to provide additional barriers to prevent nuclear wastes escaping the repository. The packaging process must be relatively simple, inexpensive, and able to handle wastes of many different compositions. The package itself must resist the heat, pressure, water, and radiation expected in a repository, as well as abnormal or unexpected conditions. The package would then be placed in a hole drilled in the floor of the repository.

The first step in packaging is to solidify the liquid wastes, which are hard to handle and transport and often are corrosive, into a powder form. The powder is mixed with some type of melted substance and poured into heavy metal canisters for eventual disposal. If spent fuel is to be disposed of, it would not have to be solidified.

Considerable research has been done to find suitable materials to mix with the powdered waste. A type of glass resembling Pyrex glass has been studied the most and was long considered the leading candidate. Then in the late 1970s, doubts were raised about this process, called vitrification. Studies indicated that the glass might break apart from the heat of radioactive decay, permitting radioactive materials to leak into the environment. Delaying disposal would allow some of the heat to escape, simplifying vitrification and permanent disposal.

Because glass has such a head start, it is likely to be the form in which wastes will be deposited in the first repository. It is possible to mix, melt, and cast glass cylinders about a half meter in diameter and 3 meters long. Large cylinders of radioactive glass were made in Great Britain a decade or more ago, and in Hanford in 1978. France is already operating a plant to bind its radioactive wastes into glass.

Ceramics, which have crystalline structures and resemble natural rock, are another possible waste form. Because they are crystalline, they are generally more stable than glass, which is not. The orderly structure of the molecules of ceramics can be tailored to bond tightly with a particular element in a waste and to cope with specific conditions of a particular repository. Like glass, ceramics require high temperatures—1,000° C or higher—in their processing.

Concrete is the third form into which radioactive wastes can be bound. Concrete has been studied in the United States and Sweden, and is being used in the Soviet Union to package wastes. Concrete can be produced at about 250° C. Because concrete is easy and cheap to manufacture, its use would enormously simplify disposal of high-level wastes. Some nuclear experts, however, question if it can hold waste for long periods of time if the package should come into contact with water.

In addition to the waste form, a series of three canisters

in a waste package is being studied by Federal scientists: the waste form canister, the overpack canister, and sleeve. They are supposed to help contain the waste for at least 1,000 years and simplify retrieving the waste, if that proves necessary.

The waste powder and melted substance, such as glass or concrete, are poured into the *waste form canister*. Its primary purpose is to preserve the structure of the waste during temporary storage, transportation, and long-term isolation. The best material for the canister found so far is stainless steel. It is strong and is not affected by heat, radiation, and other conditions in the repository. Spent fuel would be placed in the canister and packed with some kind of material to fill the empty spaces. Likely candidates are an inactive gas such as helium, or a glass or metal that would be melted and poured into the canister.

The second canister, the *overpack canister*, will be placed around the waste form canister, probably at the repository. The primary purpose of the overpack canister is to protect the wastes from any water that might enter or already be in the repository. Thus it will have to be a material, probably some kind of steel, that is highly resistant to corrosion. Federal scientists believe that the overpack canister should be able to protect the wastes from circulating water for 1,000 years. This means that even if all the other waste package materials lost their ability to isolate the waste, the overpack canister would still be able to keep the wastes separated from groundwater.

The last canister in the waste package is the *sleeve*. It,

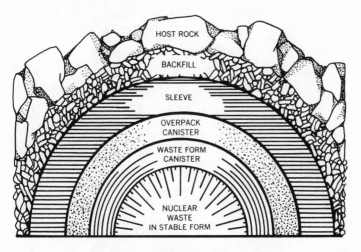

A PACKAGE FOR NUCLEAR WASTE

From the inside out, a package for nuclear waste consists of the solidified waste in a stable form that resists being decomposed, three canisters that preserve the structure of the waste and prevent water from reaching the waste and a backfill material that absorbs any water that might find its way into the repository and also absorbs any radioactive materials that escape the canisters.

too, can help preserve the structure of the waste and resist corrosion. Its primary purpose, however, is to permit the waste package to be easily retrieved, perhaps for as long as 50 years, according to current plans. The sleeve, once again probably made of steel, will be built into a repository so that the first two canisters can be inserted and removed.

The final layer of the waste package is the backfill. All the rock formations may contain traces of water. Therefore, Federal scientists are proposing to place a special material between the sleeve and the "host" rock formation to absorb the water and reduce the possibility that it will come in contact and corrode the canisters. In addition, the backfill will absorb any radioactive materials that might come in contact with water and be carried out toward the rock. The best backfill for salt is a type of clay mixed with absorbing materials such as sand and charcoal. More work needs to be done to find suitable backfill materials for other kinds of rock formations.

Federal studies made so far suggest that it will be simpler to isolate high-level wastes than spent fuel, and also cheaper, taking into account the value of the uranium and plutonium recovered from reprocessing. Burying spent fuel may be able to reduce the risk of spread of nuclear weapons for now, but it may just transfer the risk to future generations. Burying spent fuels concentrates plutonium into rich deposits that future generations may choose to mine for use in making nuclear weapons.

Federal scientists are confident that, purely from a technical point of view, geological disposal is a practical and safe way of isolating nuclear wastes for long periods of time. The only obstacle, as they see it, is the willingness of communities to accept a repository.

How Long Is Long Enough?

How long radioactive wastes remain hazardous is another issue dividing the experts. Some experts say that the wastes pose a threat only for a few hundred years; others refer to it as the "million-year" problem. Those who talk of 300 years base their belief on the average half-lives of strontium 90 and cesium 137, which is 30 years. After 10 half-lives, high-level wastes would have

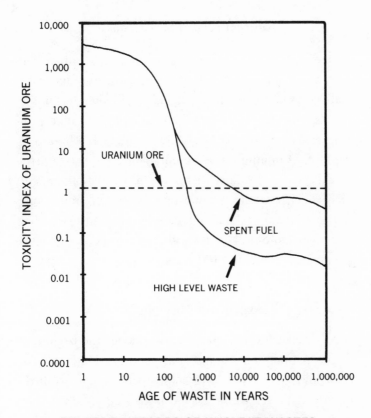

**RELATIVE HAZARDS OF NUCLEAR WASTES
COMPARED TO NATURAL URANIUM ORE**

*After about 600 years, the toxicity of high-level nuclear
wastes is equal to that of uranium ore, whereas it takes almost
10,000 years for spent fuel to decay to the same degree of
toxicity. Thus a repository must be effective for a much
longer period of time for spent fuel than for high-level waste.*

about 1,000th of the original radioactivity and no longer
would pose a significant hazard. Those proposing to isolate wastes for up to 1 million years base their estimates
on elements such as plutonium 239, which has a half-life
of 24,400 years.

119

One way of looking at the problem is to compare the relative hazards of waste buried deep underground with that of naturally occurring radioactive elements in the earth's crust, most of which have half-lives of millions of years and remain hazardous almost forever. If a hazard at the level of uranium were acceptable, reprocessed wastes would have to be isolated for about 1,000 years. Spent fuel, which contains significant amounts of transuranic elements, would take about 250,000 years to reach that level. Most low-level wastes will decay to the hazard level of uranium after about 100 years.

Transporting Nuclear Wastes

Nuclear wastes have been shipped since the beginning of the nation's nuclear program. Most shipments have been low-level wastes, plus some spent fuel. Very little liquid high-level waste has been transported from one plant to another in the United States. Transportation of radioactive materials is regulated by the Department of Transportation and the Nuclear Regulatory Commission.

Packaging requirements depend on the type of waste being shipped. Most low-level wastes require packaging similar to that used for any industrial material. Transuranic wastes are most frequently shipped in special railroad cars. High-level wastes and spent fuel must be shipped in heavily shielded casks built to withstand extreme accidents, including impact, puncture, fire, and immersion in water. Such wastes are usually shipped by rail because of the tremendous weight of the casks. So far, only spent fuel, mostly from experimental and military reactors, has been shipped routinely.

Most accidents and leakages during transportation have involved low-level wastes, and no deaths or serious injuries have occurred. In fact, the transportation of radioactive materials has a much better safety record than that for other hazardous materials, such as many chemicals. Despite this record, many state and local governments restrict transport of radioactive materials. New and stricter Federal regulations have been issued to continue to improve how radioactive wastes are transported.

The Achilles' Heel

The nuclear waste problem challenges the entire nation. It may be the Achilles' heel that will determine the future of nuclear power. Never before have engineers and scientists had to develop a tamper-proof method of isolating such large volumes of extremely dangerous materials for such long periods of time. There is little previous experience to build on, although the natural chain reaction in the mine in Oklo, Gabon, suggests it is possible. Scientists must continue to gather information needed by government officials to establish sound policies for handling and storing nuclear wastes. Not that there is any shortage of information—at a hearing before a committee of the United States Senate, a witness said "there is a rumor that the volume of paper generated in studying nuclear wastes has now exceeded the volume of nuclear wastes!" But too often in the past, the Federal Government has spent too much time in trying to justify

122

its proposals and prove that they were technically feasible, instead of studying the problem objectively.

Because management of nuclear wastes is a highly technical, as well as an emotional issue, government officials find it difficult, first to understand such issues, and then to communicate them to the public. For citizens, the challenge is to recognize that if they want to enjoy the benefits of nuclear power, they must also be willing to bear some of the responsibility for the wastes. Is it fair for states with nuclear power plants to pass laws banning the burial of nuclear wastes within their boundaries, as a number of states have done? The problem of nuclear wastes cannot be solved if everyone says bury them somewhere, "but not in my backyard." Should that happen, the Federal Government may have to override state and local opposition and select the best site available.

A cask for shipping spent nuclear fuel was subjected to crash tests by Federal scientists in Albuquerque, New Mexico. A rocket-powered locomotive (top) traveling at 131 kilometers per hour struck the cask, which was sitting across the tracks (second picture). The 25-metric-ton cask received minor surface dents during the test, but did not crack open or leak any contents. At impact, the cask was knocked into the air and bounced twice on the ground before coming to rest between the rails of the track. The front half of the locomotive was totally crushed by the impact and the trailer on which the cask was mounted was bent around the locomotive in a U-shape. The purpose of the testing program is to enable scientists to predict more accurately how well containers used to transport nuclear materials can withstand severe accidents. DEPARTMENT OF ENERGY

A Nuclear Detective Story

In the winter of 1957–58, rumors circulated in Europe that a large nuclear "accident" had occurred in the Soviet Union. Tight censorship prevented release of any information, and nothing came of the rumors for many years.

In 1976, Zhores Medvedev, an internationally known biochemist and a refugee from the Soviet Union, presented an explanation of the accident: radioactive wastes had exploded in Kyshtym in the southern Ural Mountains, the original site of the Soviet Union's nuclear reactors for producing plutonium for weapons. In developing his explanation, Medvedev did no laboratory experiments. Instead, he became a scientific detective. He used the library for research and he talked to people.

"Fortunately for a detective," Medvedev has written, "there are often witnesses to a crime. In this case, the only witnesses who could talk are those who emigrated from the Soviet Union." Following Medvedev's original article, one witness, Professor Lev Tumerman, described how he had traveled by car between Sverdlovsk and Chelyabinsk, two cities in the Urals:

"About 100 kilometers from Sverdlovsk a road sign warned drivers not to stop for the next 30 kilometers and to drive through at maximum speed.

"On both sides of the road as far as one could see the land was 'dead': no villages, no towns, only the chimneys of destroyed houses, no cultivated fields or pastures, no herds, no people . . . nothing.

"The whole country around Sverdlovsk was exceedingly 'hot.' ['Hot' in this context refers to contamination by radioactive isotopes.] An enormous area, some hundreds of square kilometers, had been laid waste, rendered useless and unproductive for a very long time, tens or perhaps hundreds of years.

"I was later told that this was the site of the famous 'Kyshtym catastrophe' in which many hundreds of people had been killed or disabled.''

All those with whom Professor Tumerman spoke during his visit, according to Medvedev, scientists as well as laypeople, had no doubt that the disaster had happened because of negligent and careless storage of nuclear waste.

Because Medvedev had to reconstruct an accident at a distance from where it occurred, he primarily used circumstantial or indirect evidence. In 1977, he began a thorough study of Soviet scientific journals that publish articles on radiation, mutations, and re-

Kyshtym, in the Ural Mountains, was the site of a large nuclear accident in the Soviet Union.

lated subjects. "The purpose of my examination was quite simple," Medvedev relates, "—to find studies that investigated large contaminated areas; that is, areas where the contamination had not been made deliberately for the purpose of a study. Obviously, a deliberately contaminated area would be small. While the authors might not indicate the location of the radioactive environment they were studying, botanical, geographical, and zoological details should make it possible to pinpoint the approximate locations. This is because some rare species of plants and animals are only found in rather restricted geographical areas."

As Medvedev began to publish his story, piece by piece, environmental activists in the United States used the Freedom of Information Act to force the Central Intelligence Agency to declassify documents relating to the disaster at the Kyshtym Nuclear Center. "Obviously, no one could be certain that the CIA had such information," Medvedev points out, "—but it did, and documents were made available to me and to other interested parties. The documents were sanitized; that is, they were 'cleaned up' by removing information that was considered still classified. Thus, more than half of the documents were not made available by the CIA. Those which were made available clearly indicated that a disaster had taken place in the Kyshtym area of the Chelyabinsk region. The documents talked about the contamination of large areas by radioactive waste. So now we have a second witness or witnesses."

Medvedev was soon joined by other scientific detectives. Nuclear scientists at Oak Ridge National Laboratory in Tennessee also concluded that a disaster had occurred. They ruled out explosions involving nuclear weapons and decided that the accident was caused by an explosive chemical reaction in a dump storing highly radioactive wastes. According to Oak Ridge, one of the methods commonly used in re-

processing spent fuel at that time produced significant amounts of ammonium nitrate, which is explosive. In the United States, wastes containing ammonium nitrate were never stored, but were treated to
destroy the ammonium nitrate.

"In addition to reviewing the Soviet literature along
the same lines that I had," Medvedev says, "the new
detectives carried out an inspection of large-scale
maps of the area prepared at different time periods.
They found that about 30 human settlements (villages and small towns) have disappeared from maps
issued after 1958, and that in the same period, a
large body of water had been isolated by a special
system of reservoirs and canals. This special system
was apparently constructed to reduce the hydrologi-
cal transport of radioactive materials through the
Techa River systems."

The dispute over what happened probably will continue for many years. But whatever happened, it appears to have been the worst nuclear accident ever.
The Soviets must have had to deal with a large area of
badly contaminated with radioactive materials. What
would be of great value to the rest of the world is why
it happened and how they have dealt with it.

VI.

Nuclear Energy at the Crossroads

NUCLEAR ENERGY is at the crossroads in the United States. Should we stop building new nuclear power plants? If we do, nuclear fission will be a transition source of energy that will see us through to the time when we have learned how to use the sun, winds, earth, or nuclear fusion to meet more of our energy needs. If attempts to develop new energy sources fail, the people may then be willing to accept the risks of nuclear energy, and more new plants will be built. Or should we continue with building new nuclear plants now? Doing so would make electricity from nuclear fission an important source of energy in the 21st century.

Electricity is a remarkably flexible form of energy. It can be converted to other forms of energy—for example heat, light, and kinetic energy—and so finds many uses in homes, businesses, and factories. In 1950, electricity accounted for only 15 percent of the energy consumed in the United States, while in the early 1980s, it accounted for about one-third.

Public Opinion

The American people are divided on the nuclear question. In 1980, 60 percent of the residents in Maine voted against a proposal that would shut down its only nuclear power plant. By the same margin, voters in Missouri voted down a proposal to prohibit the operation of two nuclear plants under construction until a repository for nuclear wastes was available. However, a proposal in Oregon similar to the one in Missouri was approved by 52 percent of the voters; no nuclear power plants were under construction in Oregon at the time.

Public opinion polls also indicate that Americans have mixed feelings about nuclear power. One poll, conducted in 1981, showed that a majority favored building more nuclear plants in the United States, and at the same time opposed having a nuclear plant within 5 miles of their community.

Other Countries

Many other nations do not share our doubts about nuclear energy. The Government of Great Britain, for instance, carried out its own investigation of the accident at Three Mile Island. Britain then decided to step up construction of nuclear power plants and placed its first order for a reactor of the same general design as the one at Three Mile Island.

In 1981, 199 nuclear plants were in operation in 21 foreign countries. These plants can generate 64 percent more electricity than all United States nuclear power plants together. Another 36 foreign plants were sched-

Nuclear power plants are increasingly the scene of
demonstrations. Here, at Seabrook, New Hampshire,
opponents and supporters turn out to express their opinions.
ATOMIC INDUSTRIAL FORUM

uled to start in 1982, and construction was underway on 115 more. After the United States, France, the Soviet Union and Japan have the largest generating capacity, while Switzerland generates the largest share—28 percent—of its electricity from nuclear energy.

Second thoughts have developed in some foreign countries, however. The governments of Sweden, West Germany, the Netherlands, and Switzerland have declared that no new nuclear power plants can be built or operated until a safe and permanent waste disposal method can be demonstrated. And in 1978, Austrians, by a narrow margin, voted against opening an already completed plant, in part because of their concern over nuclear wastes.

Countries are moving toward nuclear energy to ease their dependence on imported oil. Over 90 percent of the

Three reactors (only two shown here) are in operation at the Mihama Nuclear Power Plant in Japan, the country that ranks fourth in capacity to generate nuclear electricity.
ATOMIC INDUSTRIAL FORUM

Spain's first commercial nuclear power plant went into operation in 1971. Spain now operates four reactors and has plans for 14 more. ATOMIC INDUSTRIAL FORUM

oil consumed by countries in Western Europe is imported. Many have suffered since oil prices rose dramatically, starting in 1973. They've had to send large amounts of money abroad to buy oil, money they would rather have spent meeting other national needs. And they've felt threatened by their dependence on oil-exporting countries, some of whom have internal problems that make them unreliable suppliers.

The United States also depends heavily on foreign oil, importing 40 percent of its needs in the early 1980s, or 20 percent of all the energy it consumed. Only 30 years earlier, we were able to meet all our energy needs without imports.

The Coal Option

Unlike many countries, the United States possesses natural resources that could make us much more energy self-sufficient: enough uranium to last well into the next century and enough coal to last for several centuries.

Coal is our most abundant source of energy, but it is awkward to handle, expensive to transport, and dangerous to mine. In the 1970s, 1,547 coal miners died in mine accidents, the majority of them in underground mines. Coal mining is one of the most dangerous industries in the United States. The deaths usually occur only one or a few at a time, making it easy to forget how many lives are lost.

Uranium miners suffer fatal injuries at almost the same

Tankers carry oil across the oceans to nations around the world. Many countries want to make more nuclear electricity to reduce their dependence on foreign oil. DOE PHOTO

rate as coal miners, but there are far fewer people mining uranium than coal. In both industries, miners suffer lung diseases, coal miners from coal dust and uranium miners from radiation. In the 1970s, 69 uranium miners died from accidents and about 300 are believed to have died from lung cancer as a result of their exposure to radiation.

Mining coal at the surface, or "strip mining," is less dangerous and less expensive than underground mining. However it damages the environment, as abandoned strip mines show, although reclamation of the land is possible. About 4,000 square kilometers of United States land have been abandoned and never reclaimed.

The burning of coal pollutes the air with fine particles of unburned material that appear as black smoke and with invisible gases. Some of the gases contain sulfur, which is dangerous to health and has been involved in many major air pollution disasters. One such disaster, in London, England, in 1952, was blamed for the death of 4,000 people. Technology exists to prevent the environmental damage done by the mining and burning of coal, and many state and Federal laws are intended to prevent such damage.

Two problems resulting from the burning of coal and other fuels may not be as easy to solve. One is "acid rain." Each year, about 50 million metric tons of sulfur oxides and nitrogen oxides are discharged into the air over the United States. Through a series of complex chemical reactions, these pollutants can be converted to acids, which may return to earth in either rain or snow. This acid precipitation can damage the environment and

This giant excavator cuts away almost 3,200 metric tons of earth and rock each hour to expose coal seams that lie near the surface of the earth. The United States has huge resources of coal, which it could use instead of nuclear power to meet its future energy needs. Strip mining damages the environment, however, while underground mining is dangerous for miners. The burning of coal can also be a major source of air pollution. DOE PHOTO

destroy stone monuments and statues. Various sulfur compounds that can lead to acid rain are known to travel as far as several hundred kilometers per day while in the atmosphere. During this transport, these pollutants can easily cross geographical and political boundaries, creating national and international problems in regulating the pollutants.

A second problem is carbon dioxide, which is released into the atmosphere from the burning of coal and other fuels. In the air, carbon dioxide acts like a "one-way mirror." It lets the sun's rays reach the earth's surface, but won't let the heat escape from the earth into outer space.

135

This process is called the "greenhouse effect" because it is similar to the way glass traps heat in a greenhouse. If carbon dioxide continues to build up in the atmosphere over long periods of time, global temperatures may increase by a few degrees. Such an increase might melt ice and snow in the polar areas, thereby raising the level of the oceans to flood coastal areas, where many people now live. It might also change patterns of precipitation, bringing rain to the Sahara or turning United States corn-growing areas into dry prairies. There appears to be no immediate danger from this buildup of carbon dioxide in the atmosphere. Nevertheless, considering the vast amounts that might be added in the future from the burning of coal and other fuels, we must gain a better understanding of the possible dangers.

Some people believe that neither coal nor nuclear energy should be the answer to meeting our future energy needs. They believe that Americans are very wasteful and that consumption of energy could be reduced significantly without reducing our standard of living. Americans have already begun to reduce their energy consumption in response to higher prices. People are buying smaller cars, driving them less, and adjusting the thermostats in their homes. Industries and businesses also are finding ways to reduce the amount of energy they use.

Those who oppose expansion of conventional sources of energy also believe that we should make greater use of "renewable" sources of energy, such as the sun and wood, for which new supplies are constantly being created. Many of these sources lend themselves to small applications, such as solar-heated homes, wood-burning

Many people believe that we could meet more of our energy needs by using the sun's heat. These solar cells power an experimental irrigation project near Mead, Nebraska. The cells turn the sun's radiant energy directly into electricity. This project uses 120,000 individual cells to produce 25 kilowatts of electric power at peak sunlight. This power is used to pump water out of a reservoir and through the irrigation system. DOE PHOTO

stoves, and individual windmills. In 1977, the Federal Government for the first time began to spend significant amounts of money on finding ways to conserve energy and develop unconventional sources.

Other people believe that there is a limit to what can be accomplished by conservation and using renewable energy. In their opinion, we must continue to produce more energy, especially electricity, to meet the needs of industry and our growing population and to improve our standard of living. Under President Reagan, less Federal

A large modern wind turbine generator is being field tested on Culebra Island, Puerto Rico. The machine, which has rotor blades spanning 38 meters, is expected to provide approximately 20 percent of the energy needs of the Island during the Monday through Saturday period. On Sundays, however, it will provide all of the energy needs of Culebra, making it the only island that for a given period of time will derive all of its energy needs from renewable resources.
DEPARTMENT OF ENERGY

money is to be spent on energy, and what is spent will once again be mostly on nuclear energy.

Breeding Nuclear Fuel

To stretch supplies of uranium, the United States and a number of other countries have been working on development of a "breeder" reactor, which creates more fissionable fuel than it consumes in the process of making electricity. The breeder, which is generally similar in design to light-water reactors, splits uranium 235; in addition, the uranium 238 (which accounts for more than 99 percent of natural uranium) is converted to fissionable plutonium 239 when it captures free neutrons. Thus, the breeder can replenish its own fuel supply, as well as breed new plutonium to help fuel other reactors. A breeder recovers as much as 60 times more energy from uranium ores than a light-water reactor. Initially, the uranium 238 would probably come from wastes of the enrichment process used in preparing fuel for light-water reactors.

A number of different types of breeder reactors have been studied. Most of the research has focused on the liquid-metal fast breeder reactor in which liquid sodium serves as the coolant. This reactor has no moderator, which permits neutrons to travel at faster speeds within the reactor. All nuclear reactors create some new fuel during their operation, but the breeder is more efficient because it produces fast neutrons.

Most of the billions of dollars the United States Government spent on energy technologies until 1977 was on the development of nuclear reactors, with the goal of building a commercial-size breeder reactor by the 1990s. In 1977, President Jimmy Carter indefinitely postponed plans to build a small breeder reactor to demonstrate that the design was practical. Carter believed that the Clinch River Breeder Reactor Project, planned for a site near Oak Ridge, Tennessee, would be out-of-date and uneconomical when it was finally completed. Carter's action, which was linked with his indefinite ban on commercial reprocessing of nuclear fuel, was based on the fear that the plutonium produced by breeders would contribute to the proliferation of nuclear weapons, which have the potential for wiping out modern civilization. The United States Congress, however, kept the Clinch River Breeder Project alive, and President Reagan, in his first formal statement on nuclear energy, supported its continuation.

President Carter also organized an international study of a number of alternatives to reactors that breed plutonium. In 1980, the study, made by representatives of 66 nations, recommended continued development of breeder reactors. Other countries, including Great Britain, France, and West Germany, are reprocessing nuclear fuels on a commercial scale. Thus, Carter's two actions aimed at halting spread of nuclear weapons failed to convince other nations to do the same.

The largest breeder in the world started production in the Soviet Union in 1980. The Soviets operate four small breeder reactors and are planning a commercial size reactor. France has the most advanced breeder program in the world. It has been operating a small reactor since 1973 and, if it meets its schedule, will begin operating the first commercial-size breeder reactor in 1983. Thus, the United States, a pioneer in breeder technology, now lags behind many nations.

Benefits and Risks

At the heart of the nuclear debate is the question of benefits and risks. Nuclear energy obviously has both—as do many modern technologies. About 50,000 Americans die of automobile accidents every year, yet no one suggests banning cars. Jet planes crash occasionally, usually with the loss of many lives, but no one suggests giving up jet travel.

With nuclear energy, the question is more complicated, however, for the benefits are for society as a whole, while the risks largely are to individuals who get no special benefits—that is, individuals living near a nuclear power plant or disposal site use no more electricity than those living hundreds of miles away. The same perplexing problem arises with coal-fired power plants. Coal miners are subjected to a high risk of lung disease and accidents but use only an average amount of electricity.

Another problem is that people today are much more aware and concerned with risks than in times past. This stems, in part, from better education, which has led to increased awareness that few "miracles of modern science" are entirely free of risks. But the speed of modern communications also plays a part. Within minutes, we are able to learn about incidents taking place in other parts of the world. Often, however, we must rely on human beings to interpret what is happening and how important it is.

Supporters of nuclear energy point out that within a few weeks of the accident at Three Mile Island, ten coal miners were killed in England, seven people were killed by a gas explosion in Philadelphia, and eight oil well drillers were lost off the coast of Texas. These accidents quickly passed into history, with little or no lasting impact. The low levels of radiation released at Three Mile Island, according to a study made by Federal scientists, could cause about one extra case of cancer and one extra inherited mutation over the next few generations. But the possibility of a catastrophe at Three Mile Island, which the President's Commission considered highly improbable, was a major news story for days. It still appears in the news from time to time and will probably continue to do so until the cleanup is completed and even afterwards.

Certainties and Uncertainties

In the past few years, many people have been frightened by nuclear energy and have lost confidence in a technology that, as the President's Commission pointed out, is by its very nature potentially dangerous. Is there any way that confidence can be restored? Should it be restored? Some of the answers will come from scientists and engineers. Unfortunately, they are not in agreement now and probably never will be. Scientists are humans and do not always separate their own personal opinions and biases from facts—and many are nuclear specialists who have committed their careers to the use of nuclear energy.

The facts themselves are not necessarily known, and many uncertainties and disagreements persist. For example, information on large numbers of people over several generations is needed to decide, with certainty, how low levels of radiation affect human health. And the stability of buried nuclear wastes cannot be known definitely for hundreds if not thousands of years.

Some of the decisions about the future of nuclear

energy must be made by political leaders who are responsible for developing energy policies. Many policies require taking the long view of national needs, which may be difficult for government officials who are elected for relatively short terms of office. They (and the people who elected them) want to see the results of their decisions while they are in office. Elected officials find it especially difficult if the short-term results of their decisions may not be popular with the voters.

Elected officials are also subjected to other pressures. The companies that generate and transmit electricity make up the largest industry in the United States. It is powerful and skillful at communicating its views to local, state, and Federal officials, as well as the news media. The environmental activist groups opposing nuclear energy also are skillful. It is usually easier to frighten people than to reassure them, and the activist groups often have been effective at spelling out the risks and uncertainties surrounding nuclear energy.

In such an emotional atmosphere, the people who feel strongly about nuclear power want to label everyone as either pronuclear or antinuclear. Any attempt to examine and understand all the issues is likely to arouse suspicion. Surely the person has hidden motives and loyalties! Nevertheless, shades of opinion are emerging on both sides. Few opponents of nuclear energy favor immediately shutting down all nuclear power plants now in operation. Most oppose a specific plant or the building of all new plants.

Among some people once undecided or in favor of nuclear power, this view is emerging: from a purely technical point of view, nuclear energy is an acceptable way to produce electricity—the benefits and risks are comparable to those of other alternatives. However, according to this position, nuclear energy has aroused so many fears in so many Americans that, from a social and political point of view, it is not acceptable now. Therefore, the Federal money spent on energy, much of which is now devoted to nuclear energy, should be shifted into developing other, more acceptable alternatives.

Nuclear energy at the crossroads: Who should determine which road it should take? Americans of all ages, not special groups. The people must lead the leaders and support them in making the wise choices that will meet the nation's energy needs.

Glossary

ALPHA RAYS: a stream of positively charged helium nuclei (which consist of two neutrons and two protons bound together) emitted by a radioactive nucleus.

ATOM: the fundamental building block of the elements; cannot be divided by chemical means into simpler substances.

ATOMIC WEIGHT: the mass of an atom relative to other atoms; approximately equal to the total number of protons and neutrons in the atom's nucleus.

BACKGROUND (NATURAL) RADIATION: the radiation in the natural environment, including cosmic rays and radiation from naturally radioactive elements, both outside and inside the bodies of humans and animals.

BETA RAYS: stream of electrons emitted by a radioactive nucleus.

BOILING WATER REACTOR: light water reactor in which the water used as a moderator is allowed to boil at the normal temperature (100° Celsius at sea level).

BREEDER REACTOR: a nuclear reactor that produces fissionable fuel as well as consuming it, especially one that creates more than it consumes. The new fuel is created when neutrons from a fission reaction are absorbed by certain nuclei.

CELL: the fundamental unit of life. All living organisms are composed of cells.

Glossary

CHAIN REACTION: a fission reaction in which the products of the reaction can start a chain of similar reactions.

CONDENSER: a device that cools a gas to turn it into a liquid.

CONTAINMENT BUILDING: a concrete enclosure around a nuclear reactor intended to prevent escape of radioactive materials.

CONTROL RODS: rods made of materials that are good absorbers of neutrons; used to regulate a chain reaction in a nuclear reactor.

COOLANT: a substance circulated through a nuclear reactor to remove the heat produced by the fissioning of a nuclear fuel. Ordinary water is used in almost all United States nuclear power reactors.

CORE: the central portion of a nuclear reactor, containing the fuel.

CORRODE: to react chemically on the surface, especially of metals, by the action of water, air, or chemicals.

DECAY (RADIOACTIVE): the process by which a radioactive element spontaneously disintegrates, releasing ionizing radiation until a stable, lighter, nonradioactive element is formed.

ELECTRON: a particle that carries a negative electrical charge and spins around the nucleus of an atom.

ELEMENT: one of the 106 known chemical substances that cannot be divided into simpler substances by chemical means and from which all molecules are formed.

ENRICHMENT: the process by which the percentage of a given isotope present in a material is artificially increased to a higher percentage than that naturally found in the material. Enriched uranium, for example, contains more of the fissionable isotope uranium 235 than is found in naturally occurring uranium.

Glossary

EROSION: the natural processes, including weathering and transportation, by which an earthy or rocky material is removed from any part of the earth's surface.

FEEDWATER (SECONDARY) LOOP: the sealed system in a pressurized water reactor where water (heated by the primary loop) is converted to steam used to generate electricity.

FISSION (NUCLEAR): a form of radioactive decay in which a heavy nucleus splits in two, in the process emitting neutrons and large amounts of energy.

FISSION PRODUCTS: nuclear fragments created when a neutron strikes a nucleus and breaks it in two.

FOOD CHAIN: the pathways by which a material passes from the first organism absorbing it through plants and animals to humans.

FUEL (NUCLEAR): fissionable material used to produce energy in a nuclear reactor.

FUSION (NUCLEAR): formation of a heavier nucleus from two lighter ones, with the release of large amounts of energy.

GAMMA RAYS: ionizing radiation, similar to x rays, emitted by a radioactive nucleus.

GENE: the part of a cell that contains instructions that it needs to function and divide. The genes are duplicated and passed on to each new cell when the original cell divides.

GEOLOGY: the scientific study of the earth's crust.

GROUNDWATER: water under the earth's surface that is the source of wells and springs.

HALF-LIFE: the time required for one-half of the nuclei of a radioactive element to decay to another nuclear form.

147

HIGH-LEVEL WASTES: wastes containing such large amounts of radiation that they must be buried deep in the earth to protect humans and the environment.

ION: an atom or molecule that has lost or gained one or more electrons, making it electrically charged.

IONIZING RADIATION: any radiation that displaces electrons from atoms or molecules, thereby producing ions.

ISOTOPES: different physical forms of the same element; they differ in the number of neutrons (but not protons) in their nuclei and hence in their atomic weight.

KINETIC ENERGY: energy resulting from motion of objects; also called mechanical energy.

LIGHT-WATER REACTOR: a nuclear reactor in which ordinary water is the moderator. Two kinds are in use in the United States: pressurized water reactors and boiling water reactors.

LOW-LEVEL WASTES: wastes containing small amounts of radiation.

MILLIREM: one-thousandth of a rem, which is the unit that measures the effects of ionizing radiation on humans.

MODERATOR: a material used in a nuclear reactor to slow down neutrons quickly without absorbing many of them.

MOLECULE: a group of two or more atoms held together by chemical forces.

MUTATION: a chemical change in a gene. A mutation in a somatic cell is passed on every time the cell divides. A mutation in a reproductive cell is passed on to all cells in the offspring of the organism.

NEUTRON: a particle that carries no electrical charge and is found in the nucleus of the atom.

Glossary

NUCLEAR FUEL CYCLE: all steps involved in supplying fuel for nuclear reactors, including mining of the ore, manufacturing of fuel assemblies and their use in reactors, reprocessing of spent fuel, and disposal of wastes.

NUCLEUS (NUCLEI): the dense, positively charged core of an atom.

ORGANISM: a living plant or animal.

PRESSURIZED WATER REACTOR: light water reactor in which the water used as a moderator is kept under pressure, preventing its boiling at the normal temperature (100° Celsius at sea level).

PRIMARY LOOP: the sealed system in a pressurized water reactor that circulates water around the fuel rods in the reactor core. The heat it picks up from the reactor fuel is transferred to the feedwater (secondary) loop for conversion to steam and then electricity.

PROTON: a particle that carries a positive electrical charge and is found in the nucleus of the atom.

RADIATION: the emission of rays, wave motion, or particles from a source. Examples are visible light rays, x rays, cosmic rays, and particles smaller than atoms emitted by radioactive nuclei.

RADIOACTIVITY: The spontaneous decay or disintegration of an unstable atomic nucleus, usually accompanied by the release of ionizing radiation.

REACTOR (NUCLEAR): a device in which a chain reaction can be started, sustained, and controlled to produce heat, and from which the resulting heat can be recovered.

REACTOR VESSEL: a steel container usually surrounded by concrete and steel shields; holds the nuclear reactor's core of fuel, control rods, and circulating water.

REM: a unit that measures the effects of ionizing radiation on humans.

RENEWABLE ENERGY: energy such as from the sun, wind, and wood for which new supplies are constantly being created.

REPOSITORY: a site in a stable rock formation a few thousand meters into the earth where high-level and transuranium wastes will be buried.

REPRODUCTIVE CELLS: cells by which an organism creates off-spring.

SOLUTION: a uniform mixture of two or more different substances; commonly applied to solutions of solids in liquids.

SOMATIC CELL: cells of which the body of an organism is constructed, as opposed to reproductive cells.

SPENT FUEL: nuclear fuel that has been exposed to so much radiation that it can no longer effectively sustain a chain reaction in a nuclear reactor.

STEAM GENERATOR: a large piece of equipment for turning water into steam. In a pressurized water reactor, heat from the primary loop moves to the steam generator, where it flows through a series of small tubes. Water in the secondary loop flows around the tubes, is heated, and turns to steam.

TAILINGS: solid wastes remaining after mining or milling of ores containing metals such as uranium.

TRANSURANIUM ELEMENTS: radioactive elements that are heavier than uranium and are produced artificially.

TURBINE: a motor in which a shaft is steadily rotated by the impact of a current of steam, water, air, or other fluid upon the blades of a wheel.

X RAYS: a penetrating form of ionizing radiation emitted when a stream of electrons strikes an object.

Index